Penguin Books
Multidimensional Man

CW01456825

The central thesis of this book is that we live not in a three-dimensional but in a multidimensional world, where the exact number of dimensions is very much greater and which, incidentally, is greater for some than for others.

Ron Atkin, a professional mathematician, argues that each person inhabits this multidimensional space and, via his connections with others, helps to create a precise but flexible mathematical social structure which each individual needs in order to live a creative and imaginative life. It is in this context that he or she enjoys the magic of love, laughter and tears, and a liberation from the mockery of being an arid statistical entity.

Once these concepts are grasped – and the author is adept at translating mathematical ideas into lively idioms – the reader is free to grasp at new and rewarding insights into the 'intuitive experiences of what we humorously call life'.

Ron Atkin was born in 1926 and educated at Emmanuel College, Cambridge. He is currently Reader in Mathematics at the University of Essex. Over the past ten years he has engaged in research work into the mathematical refurbishing of the soft sciences, with support from the Social Sciences Research Council. His other publications are *Mathematics and Wave Mechanics* (1957), *Classical Dynamics* (1959), *Theoretical Electromagnetism* (1962), *Mathematical Structure in Human Affairs* (1974) and *Combinatorial Connectivities in Social Systems* (1977), as well as numerous research papers in learned journals.

Ron Atkin
Multidimensional Man

Penguin Books

Penguin Books Ltd, Harmondsworth, Middlesex, England
Penguin Books, 625 Madison Avenue, New York, New York 10022, U.S.A.
Penguin Books Australia Ltd, Ringwood, Victoria, Australia
Penguin Books Canada Ltd, 2801 John Street, Markham, Ontario, Canada L3R 1B4
Penguin Books (N.Z.) Ltd, 182–190 Wairau Road, Auckland 10, New Zealand

First published 1981

Set, printed and bound in Great Britain by
Cox & Wyman Ltd, Reading
Set in Monotype Times

to Amina, whose patience almost ran out

Contents

Preface

I have been engaged for most of my life in what is humorously known as higher education, and in that particular sphere I have been mostly concerned with mathematics and science. But in spite of that experience I have tried to keep faith with the idea that inside every scientist there is a human being trying to get out. Furthermore I am convinced that outside every scientist there is a mathematician trying to get in. So somewhere in the middle, in what might be called the human battlefield, the man and the mathematician must try to come to terms. But since these remarks might seem to imply that 'scientist' and 'human being' are antithetical (which would constitute an unacceptable paradox) I would be happier if the reader would replace 'human being' by 'artist': then I believe that the mathematician and the artist can meet in the human being. Together they can heal the wound opened up by the barren philosophy of the 'two cultures', and hopefully bring to man *qua* scientist an expression of artistic feeling for humanity – which ordinary people look for in despair.

This book is a corner of that battlefield – where the mind and the spirit are fighting it out as champions of 'form' and 'content'. Not that either side can or should expect to win, but that peace comes only when one side makes overtures to the other – and in a language which both can understand. This is my offering for such an overture, and it is couched in a 'language of structure' which, when spoken by both, allows the mathematician room for his analysis and the artist space for his synthesis. I would like only to think that it will help to bring the peace a little nearer.

RON ATKIN

Acknowledgements

I am particularly indebted to my daughter Susan, for her enthusiastic help with the analysis of the mechanics of *A Midsummer Night's Dream*; to Graham Eccles, for his teaching on Art and his colourful wit on all other matters; to Jeff Johnson, for his tireless help as a colleague and friend in fields of research where many would fear to tread; and to many friends scattered around the world – Gordon Pask, Peter Gould, John Gedye, Brian Griffiths, John Casti, David Mulhall, Moon Mullins, Dave Rosen, Lionel March, Alex Bell – who have always helped and never hindered.

My thanks are due, too, to Jill Norman, my editor at Penguins, whose encouragement has been a source of strength.

The author and the publishers are grateful for permission to reproduce the following items:

Association pour la Diffusion des Arts Graphiques et Plastiques for *Young Girl with Guitar* by G. Braque, © ADAGP, Paris, 1979

The Bodley Head for the extract 'Boarding House Geometry' from *Literary Lapses* by Stephen Leacock

Granada Publishing Ltd for the poem 'My Love' from *The Complete Poems 1913–1962* by e e cummings

Joseph Heller and Jonathan Cape Ltd for an extract from *Catch 22* by Joseph Heller

Little, Brown and Company, and The Belknap Press of Harvard University Press, for an extract from *The Complete Poems of Emily Dickinson* edited by Thomas H. Johnson: Little, Brown and Company, copyright 1929 by Martha Dickinson Bianchi, copyright © 1957 by Mary L. Hampson; The Belknap Press of Harvard University Press, copyright © 1951, 1955 by the President and Fellows of Harvard College

New Directions Publishing Corporation for the poem 'The Animal I Wanted' from *Collected Poems* by Kenneth Patchen, © 1943 Kenneth Patchen

Laurence Pollinger Ltd and Jonathan Cape Ltd for the poem 'Bedtime' from *The Sorrow Dance* by Denise Levertov

Société de la Propriété Artistique et des Dessins et Modèles for permission to reproduce *Seated Nude 1905* and *Female Nude 1910/1911* by P. Picasso, © SPADEM, Paris, 1979

Tony Young for Plates I and II

1. A Hierarchy of Cover-ups

In this book I shall be arguing the thesis that we all live in a multi-dimensional space. This idea of a dimension, it will be argued, is the precise idea of dimension that the mathematician uses. For example, we are accustomed to saying that we live in a 3-dimensional space. The Physicist and the Engineer, the Scientist in general, have all come to that conclusion or, it may well be, they have made certain assumptions in the past about dimension and space and have decided to stick with it. But without trying at this stage to criticize this point of view, we merely accept the fact that it is part of our scientific culture to talk about living in a 3-dimensional space. After all, as we learn at school, we all can sense the difference between a line (what is called a one-dimensional thing), a plane (a two-dimensional thing), and a volume (a three-dimensional thing). These things strike us intuitively as being qualitatively different. In order to give a precise quantitative or mathematical idea to the difference the ancient Greeks certainly produced a kind of geometrical language which had this difference built in, and the way it was built in has been accepted by the modern mind as a plausible and adequate distinction between the three ideas. Hence we are educated at school and throughout our lives with this kind of cultural assumption. It is a scientific assumption. It is part of our scientific and technological equipment.

Incidentally it is of great interest in this respect that now in 1979 the scientific community throughout the world has had the benefit of the experience of all that talk earlier in the century of a *four*-dimensional world. This idea is primarily associated with the name Einstein and at the technical level with the theories of relativity which he propounded in the early part of this century. This led to a great deal of serious and not-so-serious discussion of the concept of four dimensions. The general theories put forward by Einstein and elaborated by others were precise in their use of this idea of four dimensions; in fact the fourth dimension became something intuitively associated with our concept of time – or better still with the strange and yet persistent concept of time which the scientist in his laboratory had used for two or three hundred years. This fourth dimension had to be a measurable thing in mathematical terms and so it had to look like the same sort of thing as the measurement of a line or an area or a volume. In this book we shall not need to pursue these very sophisticated technical details, but here it is highly relevant to notice that phenomenon in the intellectual thinking of this modern age. Hence it is within the normal repertoire of a scientifically educated man to use the words 'a four-dimensional world'. If he is trying to make a clever point, either in debate or at his favourite cocktail party, he will be able to take exception to the statement that we are living in a three-dimensional world and be able to show his erudition and scholarly pretensions by correcting it to the statement: we are living in a four-dimensional world.

But the thesis of this book is that we are in fact living in a multi-dimensional world where the precise number of dimensions is very great. We shall argue that it is greater for some than for others and in this respect there will be echoes of that classic quotation from Orwell's book *Animal Farm* in which he said 'all animals are equal but some are more equal than others'. But we shall argue that every man lives in a multidimensional space and with his fellows helps to create a precise mathematical structure in which they may live together. It will not be the case that any particular individual will be trapped at some peculiar dimensional level but rather that each of us, depending upon his energies and his imagination, will be able to change the dimensional structure in which he lives. Indeed, it will probably be arguable that this ability to change the multidimensional structure is a fascinating peculiarity of what might be loosely called living organisms. With that kind of view it will be natural to see the rather rigid three-dimensional space of the scientist as something which is unchangeable because it is peculiarly associated with the inanimate universe – or better still, if we understand that the universe is a three-dimensional structure per-haps we are merely insisting that the universe is inanimate!

But, you may well say, what kind of a crazy idea is this, that we actually live in a bizarre multidimensional space? Will it not be the case that it will be such a complicated kind of world to live in that we cannot ever understand it? Will it not be the case that to attempt to describe such a world in a meaningful way, that is to say, in a way which allows us to play a rational part in it and to understand it in a certain rational way, the mathematical symbolism and the mathematical ideas will be so complicated and abstract that, at the very best, a mere handful of people will be able to understand them, and would it not therefore be largely useless, abstract, esoteric and sterile? Well these questions are so big and so frightening that they cannot be answered by a simple Yes or No at this stage. After all, this is merely the first chapter of the book. The aim of this discussion is to first explain how these concepts of multi-dimension can be properly described and how it is that we might well be living in such a space without having rationally and consciously known it, and that the consequences of asserting this proposition will in fact give us a new insight into our intuitive experiences of what we humorously call life.

To begin with I wish to keep reasonably clear of a lot of mathematical talk, and indeed it will be important to keep that talk down to a minimum. I wish to pursue certain ideas which lie underneath a great deal of our intuitive experience of the world and which are expressed in the languages we use to describe those experiences. But we cannot think about the relationships between people and between people and things except by using concepts which are expressible in language – sometimes that language is a normal idiomatic thing like English, French, Chinese, etc., and other times it becomes a kind of specialized thing like the language used by the Engineer or the Chemist or the Mathematician. However, the structure of these languages is itself an expression of our intuitive feelings of the structure of our experiences, whether those experiences are commonsense data (such as the Physicist explores at certain levels in his laboratory), or whether they are such sophisticated collections of sets of data that we are not sure whether they are outside us or inside us. Is then the work of the Artist an expression of this experiencing of the Universe? Presumably it is. Is not the work of the Poet an attempt to express in language a very sophisticated view of the Universal experience? Presumably it is. I shall argue that the basic ideas of dimension and structure for which we are all searching, and trying to express in our various ways, are most precisely understood and recognized when the language we choose to talk about our experiences is essentially mathematical. I know that this might well require allaying the fears of many readers for whom the mere mention of the word mathematics is an inducement to wild hysteria.

But with all respect to our mentors I must stress that my own belief is that the experience which leads to this reaction, an experience of being taught mathematics, must have been one which completely misses the point of what mathematics is all about. Indeed mathematics is not really about anything in particular, it is merely a *language* which may be used to talk about things. Thus it has no absolute aim in what it is to talk about. For example, Algebra is not about Arithmetic but is a language which allows us to talk about Arithmetic and because of this it allows us, for example, to talk about different kinds of Arithmetic and therefore to invent different kinds of Arithmetic. If mathematics is a language then presumably it has been developed by the human intellect in order to find a better expression for certain kinds of experience than can other languages such as English. This better expression, when the mathematics is actually applied to some area of experience, is one which helps to eliminate ambiguity and to isolate delicate nuances of meaning – albeit in a restricted universe of discourse – than do the idiomatic languages. But in spite of all this, even if the final expression of structural ideas is to be found by using a mathematical language it must presumably be true – if this thesis is to be maintained – that our normal language contains in it the same sort of concepts of many dimensions and of structure which the mathematical description will eventually expose. It is therefore important to illustrate the basic concepts which are ultimately to dominate the discussion, by finding strong hints of them in the use of our everyday language and in the everyday experiences we have, both of men and of things.

Of course one does not write a book of this kind from the front to the back, one writes it virtually from the back to the front, and that is because it can only be the result of many years of groping about in the dark – in my own case by being absorbed in research work of various kinds at the University, and of trying to bring order out of apparent chaos. I say all this because when we begin to lay out some of the basic concepts the first one or two must be taken on trust. They are introduced because they are to play a profound part in the sequel. They are not the result of some random guesswork, nor are they supposed to be immediately and obviously profound, that is to say they can be greeted with suspicion in the first instance because if that were not the case, it would not have been necessary to spend so many years searching for them.

A precise hierarchy

The first of these ideas and an extremely basic one, perhaps even the most important, is the idea which I shall describe as 'a hierarchy of levels'. The use of the word hierarchy is immediately dangerous because it has frequently been used over many centuries to encompass certain rather rigid ideas. The most important of these is the idea of caste, and hierarchy has often been rather like the idea of a caste when it has been supposed to mean a division into ranks or levels of excellence or of authority or of tradition or of custom or of military responsibility. However, in spite of this traditional use of the word I have not been able to come up with a better one and so I am burdened with the responsibility of first explaining why this word, when I use it, does not mean what it has always meant in spite of there being some resemblance. The easiest thing to do is to illustrate what I mean by spelling out some very easy examples. It is as simple as this.

The word GARDEN expresses a concept which *includes* the concepts expressed by the words, FLOWER, SHRUB, TREE, LAWN, etc. Because of this *inclusion* we shall say that there are two distinct hierarchical levels involved. They may be shown schematically as follows:

Level-2   {GARDEN}
Level-1   {FLOWER, SHRUB, TREE, LAWN}

The process can clearly be continued. For example, the word FLOWER expresses the concept which includes other concepts such as those expressed by the words ROSE, DAFFODIL, HYACINTH, etc. Similarly for the word SHRUB we might well wish to include a whole set of words such as LILAC, AZALEA, JUNIPER, etc., whilst for the word TREE we can naturally make a list of familiar specific trees such as ASH, CHESTNUT, ELM, and so on. Thus we obtain an extended set of hierarchical levels which are based upon the idea of one level somehow including the ideas of the level below it.

2   {GARDEN}
1   {FLOWER, SHRUB, TREE, LAWN}
0   {ROSE, DAFFODIL, LILAC, AZALEA, ASH ...}

Then for example at level-1 the word FLOWER is really a name for a whole set of words {ROSE, DAFFODIL . . .} taken out of level 0. At any stage or level we expect the list of words to be finite. Our use of dots merely indicates laziness, and by patience and diligence we must be able to produce a complete list of suitable words at any one level. Of course some immediate questions will no doubt spring to mind.

1. Must the words all be nouns?

    We answer No to this (but see the following chapters).

2. How can you be sure that the set of words at any one level is complete?

    We answer this by imagining that we are discussing it with an *opponent* (not to be confused with the word *enemy*) who challenges the rest. If the opponent can produce additional words at any one level we gladly accept the additions. Thus ultimately all the participants in the discussion must agree about the lists. This is not such a difficult feat as one might imagine in any one particular field of study, or for a specific purpose. For instance, the lists which Physicists and Chemists classically agree about do not cause profound difficulties.

3. Why do you call the levels 0, 1, 2? What is the significance of saying the 0-level?

    The numbers at the levels are not in any sense *absolute*, and the 0-level is quite arbitrary. Of course if we want to go further down the hierarchy below the 0-level, then we either call the next one (-1)-level or we call it 0-level and rename the others, 1, 2, 3. The mathematician will get round this 'floating origin' problem (which is a trivial one of description) by calling one level the N-level (N for 'normal' or 'neutral' serving as a temporary origin) and then going up the hierarchy with $(N + 1)$, $(N + 2)$, $(N + 3)$, etc., and down the hierarchy with $(N - 1)$, $(N - 2)$, $(N - 3)$, etc.

4. Can it happen that two words at the $(N + 1)$-level have an *overlap* at the N-level? Does that cause trouble through the ambiguities involved?

    It can certainly happen that there is an overlap as described. It does not cause trouble as such but rather indicates how the language (which is English) is structured by some sort of connectivity being manifest among the concepts of discourse. This overlap can only be handled precisely and made use of to our advantage by the introduction of our later mathematical language. In the meantime we illustrate this overlap by adding the word PARK at the level $(N + 2)$:

$N + 2$ {GARDEN, PARK}
$N + 1$ {FLOWER, SHRUB, TREE, LAWN...}
N {ROSE, DAFFODIL, ... LILAC, AZALEA, ...
          OAK, ASH ...}

Now with these particular lists at $(N + 2)$ and $(N + 1)$ the words GARDEN and PARK overlap. They overlap completely if we assume that each concept includes only the four words at the next level $(N + 1)$. In this case there will be no need to have two words at $(N + 2)$. The word GARDEN and the word PARK will be synonyms. The fact that we know that normally they are not synonyms suggests that at the $(N + 1)$-level there are more words which when added in will introduce the distinction between GARDEN and PARK. For example, we might need a word at $(N + 1)$ which indicates the size of the GARDEN, or whether it is private property. However, no matter what extra words are introduced at $(N + 1)$, the words GARDEN and PARK at $(N + 2)$ will clearly have a decided overlap because of their common elements at $(N + 1)$. In the same way if we go down from the level $(N + 1)$ to the level N we might well find that there is again an overlap demonstrated between some of the words. For example, we might decide a LILAC SHRUB might just as well be treated as a LILAC TREE in which case the words SHRUB and TREE certainly overlap, by having in them at the N-level the word LILAC. Clearly this kind of thing will go on and can lead to what is apparently quite a complicated story of things being mixed up. An important point of our discussion is that the mix-up or overlap as one moves from one level to the next is not something to be decried, to be afraid of, to try to avoid, or to try to rub out; it is an important part of the structure of the language – that structure which has inherently expressed our intuitive need to have things connected to each other. What we shall do in this book is to pursue this overlap and, instead of trying to eliminate it, try rather to describe it in as precise a way as possible so as to see its implications and make full use of its virtues.

All this rather supports the idea that we might well refer to some concepts as 'bigger' than other concepts and some words as 'bigger' than others. But if we do this we must insist that we are only using the word 'bigger' by reference to the hierarchical levels which we have specifically demonstrated. For example, in the above example, the word GARDEN is an $(N + 1)$-level word compared with the N-level word ROSE. Only in this sense might we then allow ourselves to say 'GARDEN' is a bigger concept (or a bigger word) than 'ROSE'.

Big words (high hierarchical-level words) are very useful and also very limiting. Generalities are expressed in big words; for example

Beauty is truth, truth beauty

is saying something like 'beauty' and 'truth' are at the same level, say N-level, and their overlap is complete (they are identical) at the $(N-1)$-level. As in our example above it is like saying

Garden is park, park garden

at the $(N + 2)$-level. It is clear that big words are the stock-in-trade of politicians and public speakers.

The greatest happiness for the greatest number

leads to, say, the $(N + 3)$-level word 'happiness'. This safely leaves the meaning (or the significance) of the word to the listeners as that significance which is manifest to individuals at lower levels, say $(N + 1)$ or N-level.

The Nation wants peace, not war

is full of big words, heady, referring to 'nation' at a high level as in the following:

$N + 3$	{NATION}
$N + 2$	{ETHNIC GROUPS, ECONOMIC CLASSES, MALES, FEMALES...}
$N + 1$	{CITY DWELLERS, FARMERS, FACTORY WORKERS, LABOUR, MANAGEMENT...}
N	{FRED SMITH, JOHN PERKINS, CARL SCHULTS...}

In England (1979) the Planning Authorities have recently tried to include the citizens in their decision-making by an appeal for public participation. Thus if an authority is about to reorganize a regional area they are inclined to send out questionnaires to interested citizens and to ask such questions as:

1. What kind of region do you want for yourself and your children?
    (Impossible to answer, generally)
2. Do you want open spaces, better transport, fine parks?
    or concentrated housing with wide open countryside?
    or high-density industry with low-density residence?

Why, we might cynically ask, do they not go for the biggest words all at once and simply say

Do you want to be Happy in the New Plan?

Whatever the citizen answers (and usually very few bother to answer at all, sensing the futility of it) he wakes up one morning with a terrible shock as the bulldozer rips through his front garden on the first leg of the New Plan. 'That's not what I meant', he cries in anguish. Of course not; he meant something at his own sensible experience of levels, but he is in danger of getting an unpredictable plan, and one which even the planners cannot predict since their thinking is fixed at a higher hierarchical level.

$N + 3$ Big-word level: {HAPPINESS, FINE SPACES, PARKS, RESIDENCES...}

$N - 2$ Real-life level: {STREET CORNER, SHOP, MY GARDEN, TRAFFIC...}

But just as big words have an appeal for politicians and men-of-affairs, so we must admit they attract all of us to some degree or other. Much undisciplined debate (the kind which is commonly called argument) consists of two opponents hurling big words at each other, fine phrases which cannot be separated because of the severe overlap between them. Phrases such as

> I believe in justice for all
> Women are terrible drivers
> Male Chauvinist Pigs
> The workers are lazy and greedy
> You must show Christian love to your fellow men
> We need more self-sacrifice from everybody

and similar phrases. These all illustrate possible varieties of hierarchical levels in a universe of discourse containing many kinds of sets.

When Socrates said 'All men are mortal' he was referring to a hierarchy of two levels:

$N + 1$ {MORTAL-BEINGS}
N {MEN, ANIMALS, PLANTS, ...}

and since the $(N + 1)$-level is a set containing a single concept it *must include* each of the concepts in the next N-level. Socrates is merely asserting that the word 'MEN' is to be found in this hierarchy at the N-level, and at the same time he is asserting that at the $(N + 1)$-level the set of concepts only contains one.

The idea of a cover set

We have already sneaked in a certain mathematical idea by using this rather crafty N when speaking of the N-level. Now we must introduce a new word which is not merely an invention of the mathematician but is a homely word, and the mathematical use of it does not contradict the ordinary idea of what it means. This word is the word *set*. In looking at our lists of words at various levels we have in fact been looking at sets of things. In this book the word *set* will mean a list of things. We shall try to show that sometimes a set which contains certain words ought not to contain other words, but these are details of how best to organize words and ideas into sets. The word *set* itself is to mean a list of things and we may suppose that the list has been agreed upon by the reader and his opponents – sitting down together and coming to agreement by throwing words back and forth between them. Thus we are really talking about *sets of concepts*, and these become translated or represented by sets of words, specified at different hierarchical levels. But we immediately need another word which mathematicians make use of and which will summarize what we mean by the difference between hierarchical levels. This next word is *cover*. The second set at $(N + 2)$, namely {GARDEN, PARK}, will be said to be a *cover* of the set at $(N + 1)$ {FLOWERS, PLANTS, ...}. This word *cover* therefore includes that meaning which is very much like the homely meaning – as covering something with something else. The Piecrust covers various things inside the Pie. What we have in this particular analysis of ours is a whole collection of Piecrusts, and these are said to be things at the $(N + 1)$-level, while all the Plums and

Apples and Meat, etc., which may be found in the Pie are things to be found at the N-level. There is one detail which we must take care of and that is that the set of words at the $(N + 1)$-level must completely cover the set of words at the N-level. We cannot have some odd word at the N-level sticking out like a sore thumb and not being underneath anything. Hence the mathematician says that a cover of a set is another set (at our next level) and that this other set really consists of words which act as meanings of *subsets* of the first set, it being always understood that no element of the first set escapes from being covered by something in the second set. Now this word *cover* really contains the whole essence of what we mean by moving up or down a hierarchy of sets of ideas, and all our intuitive experience as expressed in our idiomatic language seems to be an elegant illustration of this very idea.

We shall generally expect a cover set at $(N + 1)$ to be seen to overlap at the N-level – it is as if the various piecrusts overlapped the various pies underneath and any one crust will be seen to be half a crust in one dish and half a crust in another. As we might expect, we do not regard this as a disadvantage but welcome it as a sign of significance in our intuitive awareness of structure.

Of course there will be special cases which arise in which the elements at $(N + 1)$, being elements of a cover-set of a set at N, do *not* in fact overlap. When this happens we have what is traditionally called a *partition* of the N-level set by the $(N + 1)$-level set. This idea of partition is deeply ingrained in the scientific attitude of the physical scientist.[1] It is also deeply ingrained in the traditional uses of the word hierarchy. This is why it was necessary to give warning of the subtle difference in *our* use of the word hierarchy and yet to admit to not being able to find a better word to describe the idea. The traditional hierarchy is actually the idea of a series of levels of sets in which, for example, the $(N + 1)$-level set acts as a partition of the N-level set below it, and so on down the whole hierarchy. This is illustrated, for example, by Koestler in his book, *The Ghost in the Machine*.[2]

There, he says: 'Thus the complex fabric of social life can be dissected into a variety of hierarchic scaffoldings as Anatomists dissect muscles, nerves and other correlated structures from the pulpic mass.' This dissection is in fact a partition, and Koestler illustrates the idea by drawing it – what the mathematicians call a *tree* – and it looks something like the following:

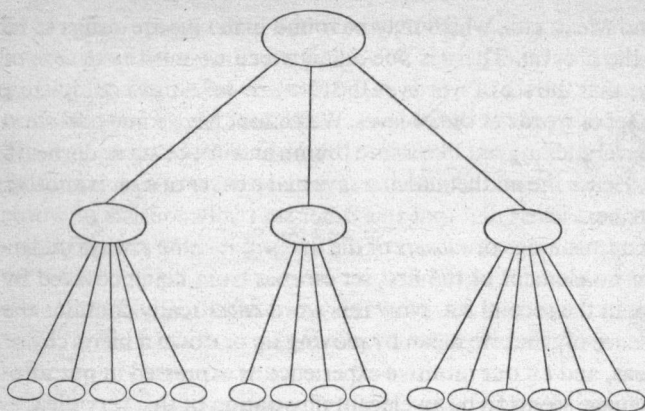

Fig. 1   Old-fashioned hierarchy based on partitions

Here you will see that the middle-level things act as a cover for the bottom level of things but are themselves quite disjoint – they do not overlap. This is an easy illustration of the traditional attitude towards hierarchy. The same things occur in the sort of tree-type structure of a military organization, where the 'chain of command' consists of moving down the hierarchy level along paths which end up at the bottom of the pyramid, with a partition of the lowest units. In the same way the word caste is normally seen as denoting a partition of society into separate pieces which have no overlap. It is often a misfortune for an individual if he finds himself in one of these pieces and, because there is no overlap, it is impossible for him to move on to any other piece. In the class-ridden society of the English industrial scene it is generally regarded as significant that a worker on the factory floor, turning the nuts and bolts, is in a class which separates him completely from the managerial individual filing memoranda in his office. This, if it is a reality, is an illustration of a partition of all the people into at least two subsets. Because there is no overlap it is impossible to move from one to the other. In comparison it has often been said that in American society every poor farmboy can see himself as President of the United States. This would suggest that traditional ideas of freedom, liberty and equality are closely bound up with ideas of structure and of overlap between the elements in a hierarchical set of covers. The real intuitive awareness of what it is like to be trapped in a partition, as opposed to a more general cover, would be found by being either a serf or a baron in a medieval feudal society. The traditional use of the word hierarchy, as a sort of caste system, is a use which insists that the individual man finds himself at some particular N-level and that there is an impassable barrier between that level and the one above it, $(N + 1)$. This impassable barrier is a social obstacle, and one which might often be enforced by violent means.

In contrast to this traditional idea of hierarchy, based on partitioning, we now search for something more general, for a view which is not dependent on partitioning (while not excluding it) and which sees society, for example, as containing a kind of hierarchical structure open to Everyman. The first kind of hierarchy – a 'chain of command' kind in which each man is slotted in at some (usually) fixed level – is illustrated in our long history of power-seeking, of castes and of divisive social classes. But perhaps we can regard that history as an interesting corruption of something more natural and unavoidable, of a kind of 'hierarchy of concepts' which is altogether more acceptable?

Imagine some 'land of innocents' busy with simple agriculture and crafts, not burdened with 'chains of command' in any sphere. What kind of *concepts* would we expect to arise in this society, what set or sets of ideas which, without being sufficient, are nevertheless necessary? Even if we only think of the primitive economics involved we can see the need for sets like the following:

$N + 2$ concepts of {this society, other societies}

$N + 1$ concepts of {manufactured goods, foods, livestock, trading services, ...}

N concepts of {carpentry, pottery, husbandry, building, farming, ...}

$N - 1$ concepts of {woods, clays, metal ores, animals, seeds, minerals, ...}

The concepts of $(N + 1)$ act as a cover (not necessarily a partition) of those at N, and so on. What of the individual who is intellectually sensitive to the $(N + 2)$-level concepts? Should he not be elected to think about the social implications at this level, to concern himself with relations between 'this society' and 'other societies'? What of the individuals who are intellectually sensitive to the $(N + 1)$-level concepts, who can understand the subtle connections between those N-level ideas which result in the successful activity of husbandry or of farming? Should they not be encouraged to accept some sort of 'responsibility' for the rational conduct of these $(N + 1)$-level affairs?

Is it not feasible that the hierarchical levels of concepts could give rise to a sort of hierarchy of responsibilities, to special categories, to special social titles? These levels will then give rise to positions of 'power' – the word which will replace, and get confused with, 'responsibility'. If this new hierarchy becomes ingrained, by the neglect of 'content' in favour of 'form', then the traditional 'chain of command' hierarchy is well on the way to being established. Unfortunately, of course, the individuals in our 'society of innocents' might all be capable of existing at the intellectual concept-levels of $(N - 1)$, N, $(N + 1)$, and $(N + 2)$. Once the 'form' is dominant over the 'content', regarding these concept-levels, then the struggle for 'power' (for the *sole* responsibility for an $(N + r)$-level) has replaced the innocent acceptance of (what was once called) responsibilities.

In this book we are, first and foremost, concerned with the underlying hierarchy of concept-levels. These levels are normally associated with cover-sets, rather than with strict partitions, and in a particular community denote intellectual conditions which *all* men might feel able to achieve. This scheme of cover-sets has a representation more like the following:

Fig. 2   Hierarchy based on realistic cover-sets

But all this hierarchical theory is not only theory about *individual* people and the way they organize themselves in a precise structural sense into what we think of as societies. The hierarchical structures we are talking about are structures of ideas as well as of people; they are also structures of physical space as well as of people. Thus we have a natural hierarchical view of geography. We speak of the level of the country, then of the region, then of the area, then of the town, then of the street, then of the house. This is already a hierarchical structure of six levels, and if we replace each level by a separate word, for example a list of all towns which we have in mind, then this is normally an example of a cover between any two neighbouring sets which is also a partition. Thus we have learnt from experience that the two counties of Yorkshire and Lancashire must have non-overlapping regions in their make-up. The territorial fighting which is supposed to decide issues of boundaries can be seen as attempts to change the non-partitioning cover into a partitioning cover. This formal and mathematical aspect of the problem already points to a study of the authoritarianism of social organization. An individual man living in Yorkshire might conceivably have a certain dual loyalty both to Yorkshire and to Lancashire (he might have accidentally married a Lancashire lass). In this case the words Yorkshire and Lancashire already overlap in certain areas of human experience, through the people they share in this way. Perhaps, too, if it is possible for them to share people it is possible for them to share ideas – of course looking *down* the hierarchy. If we look *up* the hierarchy, then Yorkshire and Lancashire must be part of a super cover at the next level. This super cover might be the concept known as 'the North of England' or it might be the concept known as 'England'. It is pretty clear therefore that moving up the hierarchy through a series of cover-sets automatically brings things together by making them less detailed, whilst moving down the hierarchy emphasizes not only the things they share but also any differences they might possess.

In those situations in which we try to be precise in our discussion of any particular system of hierarchies and sets we shall not only need complete lists of all the entities in all the sets at each of the hierarchical levels but also we shall need the precise relationships between each hierarchical level and the one beneath it. A typical relationship will tell us exactly how an $(N + 1)$-level set of entities acts as a cover of an N-level set of entities. Knowing this precise relationship, we shall then be able to analyse how things are connected and how they differ. This means that we shall need to examine very carefully, that is to say from a mathematical point of view, just what the nature of a relationship really is. The details of this kind of examination will be laid out in future chapters. For the moment we continue our study of the occurrence of these hierarchical levels in different areas of human experience, and will illustrate that much important discussion can be less precise than the mathematical analysis normally requires.

Semantics and hierarchy

In that very instructive book *Semantics* by Geoffrey Leech, he gives an entertaining account of how his two-year-old daughter began to learn English.[3] During this process she clearly developed a hierarchical view of the language. In fact she appeared to learn the language by coming down the hierarchical level. That is to say, she first seemed to master certain words, some of which were invented by herself, which acted as a cover for many other words which she would presumably learn later on. Leech gives an interesting little list of some of these words in the chapter where he discusses the idea of conceptual categories. Suffice to say that these conceptual categories are, in our terms, simple examples of higher levels in the hierarchy of concepts.

Word used	*Apparent range of reference*
choo-choo (i.e. toy engine)	toy with wheels
ba-ba (i.e. sheep)	animals in fields (including cows)
book	books, leaflets and newspapers
soo (i.e. shoe)	footwear generally, including boots and slippers
cup	vessels for drinking, including mugs, glasses, and the cat's bowl
Tom (her brother)	boys in general
on (i.e. orange)	yellow as well as orange

If we put the words in the left-hand column at an $(N + 1)$-level then an extensive list of words in the right-hand column would automatically be a set of words at the N-level. Thus the left-hand words, the so-called baby talk, act as a natural cover for this extensive list in the right-hand column. This would suggest very strongly that the learning process, as far as ordinary language is concerned, is a process which takes us down the hierarchical levels. This seems an important point in the study of semantics and it seems also to be important in the study of how the human intellect copes with its experience of the outside world. It seems pretty clear that only by starting with big words, in this special sense of the word big, can we cope with what is clearly an extraordinarily complicated set of impressions which come to us from outside experience. The child's experience of learning language seems to be a nice illustration of this process. It starts at, let us say, an $(N + 1)$-level and moves down the hierarchy as it replaces these 'big' words by subsets of lists of words and of ideas at the N-level. This process is presumably repeated as he matures and extends his vocabulary and his set of concepts by moving down the hierarchy yet further into $(N - 1)$, $(N - 2)$, etc.

The problem of learning a language as a phonetic experience and as a semantic experience is obviously conceptually associated with, but presumably is not identical with, the experience of learning how to write a language. In writing a language we are clearly moving *down* the hierarchy and ending with a basic set of signs which we call the Alphabet. This means in effect that we end up with a very large set of a very large number of each of the letters of the alphabet. In writing words in the language we are then selecting a collection of these letters of the alphabet, collecting a subset of this low-order set, and by writing them down in a certain sequence we are creating words which presumably exist at higher hierarchical levels. Thus the process of writing and presumably of learning to write is the process of starting at a very low hierarchical level and moving upwards. Is it possible for us to do this without first having the experience of moving down the hierarchy when we are learning the semantics of our language in our baby talk?

Hierarchy in chess playing

The reader who knows nothing about the game of chess might be well advised to skip the next few pages – he will be missing only another example of the occurrence and significance of hierarchical levels. But chess players know that this is not just another game with a few fancy rules, to be played by a computer (usually not very well, incidentally) as it might play noughts-and-crosses (what the Americans call tic-tac-toe). To them, and certainly to the Chess Masters and Grandmasters, this Royal Game is an activity which searches deep down into the soul. It is profound and difficult, sometimes seeming to be infinite in its possibilities and, as the great Emmanuel Lasker once said: '... on the chess board lies and hypocrisy do not last long ...' The man who plays chess at all seriously must learn to live with himself and all his weaknesses. If he is greedy then he will not easily be able to resist grabbing a pawn or a piece – even if his opponent has gratuitously offered it to him, sensing his opponent's greed. If he is self-indulgent and full of fantasies, he will be guided by wishful thinking and often not see the possibilities he is offering his opponent. If he is cowardly and timid, he will hesitate too long before throwing all his forces into an attack – which he will not have properly prepared. If he is too aggressive, he will attack prematurely, probably from the first move, and fail to appreciate that there is anything subtle about the structure and conduct of the game. If he is a reckless spendthrift, he will be reckless in sacrificing his precious pieces and be shallow in his appreciation of his opponent's intentions. And if he is acutely egotistic, he will sneer at his opponent's abilities and convince himself of his invincibility – even up to the very point of his débâcle (such players never know when to resign with good grace). Playing chess is like illness, a lonely experience. As we shall see later on, it transcends other, simpler games by the dimensionality of its structure – which we shall specify precisely.

But to return to the notion of hierarchy in the game, we shall see that the way Masters and Grandmasters play illustrates the differences between what is commonly called *tactical* chess and *positional* chess. At the so-called tactical level of play the aim is to make moves with one's pieces in such a way as to achieve some limited and rather simple-minded end. Such an end might well be nothing more than to capture a piece or to check the opposing King. But the serious player of chess knows that these kinds of tactical moves must usually be subordinated to a much higher level of appreciation of what the game is all about. In order to express this latter appreciation writers on the subject have developed a kind of language about the positional play in the game. This language uses such words as 'King-side', 'Queen-side', the 'Centre of the board', or the 'seventh rank' (useful for deploying the rooks), or the 'king's file', etc. These words express collections of squares on the board (*sets* of them) and therefore they must exist at a level which is hierarchically higher than do the words which describe individual squares on the board. A chess Master brings the same sort of hierarchical thinking to dealing with the pieces. Thus he will think about 'the pawns' as a new single entity, or about the 'heavy pieces' (usually meaning the Queen together with the Rooks), or the 'minor pieces' (the Knights and the Bishops). On the other hand he might interpose an additional hierarchical level and think for example of 'the King-side pawns' or 'the Bishops', and then he is really using a hierarchy of three levels. When using three levels he will then cover the squares by two extra levels, using such ideas as 'the King's Bishop File' (for all the squares which lie in the vertical line), before reaching the next level in which that idea is subsumed under the general 'King-side'. In general we can understand a great deal about Master play by appreciating only these three possible hierarchical levels:

Level	Nature of play
$N + 2$	Super-positional play
$N + 1$	Positional play
N	Tactical play

At the N-level the Master needs to contemplate individual moves solely from the relation between individual pieces (or pawns) and individual squares on the board. This is the fine detail of the game and is involved in the final decision about every move. But at the $(N + 1)$-level he is thinking about the effect of moving sets of pieces (the Bishops, Knights, Rooks or Queen, or King-side pawns, etc.) on to sets of squares (such as the Centre, the Queen-side files, or the seventh rank, etc.). At the $(N + 2)$-level he is thinking about even bigger sets of pieces (such as all those he might be able to use in a combined assault, or all the minor pieces, or all four centre-pawns, etc.) on broader areas of the board (such as the enemy King-area, the whole of the King-side, etc.). These successive levels will define a hierarchy of cover-sets (not normally a partition) and it is against this hierarchical background that the thinking about the game goes on. This thinking will include such well-known features as 'gaining control' of squares (N), files (N + 1), diagonals (N + 1), ranks (N + 1), King-side (N + 2), Queen-side (N + 2), Enemy King-area (N + 2), Centre area (N + 2), Defence area (N + 2), etc. Precise classification of these levels and descriptions of how the whole process can be computerized will be found in references to current research work undertaken by the author.[4,5] With such a hierarchy the Master can cope with questions like the following:

1. What positional features must be used in the opening play?
2. Does the position justify a King-side attack?
3. Does the position justify material sacrifice?
4. Should White attack or defend?

Since the game of chess is often regarded as an intriguing illustration of how the human mind copes with problems and their possible solutions, it is not surprising that a clear understanding of the game and the way it is played by humans (*not* to be confused with the way which computers are normally made to play the game) has become an important problem for researchers in the field of artificial (or machine) intelligence.

This kind of analysis, but without some of the computer and mathematical details which play quite an important part therein, gives us a means of describing a typical Grandmaster game, that between the then world champion, Emmanuel Lasker, and Napier at Cambridge Springs in 1904. We list the game here, in standard English notation, and describe the play relevant to the positions reached at the indicated points.

LASKER (White)	NAPIER (Black)	*See notes for relevance to hierarchy*
(1) P–K4	P–QB4	
(2) N–QB3	N–QB3	[a]
(3) N–B3	P–KN3	
(4) P–Q4	P × P	
(5) N × P	B–N2	
(6) B–K3	P–Q3	[b]
(7) P–KR3	N–B3	
(8) P–KN4	Castles	[c]
(9) P–N5	N–K1	
(10) P–KR4	N–B2	[d]
(11) P–B4	P–K4	
(12) N (Q4)–K2	P–Q4	[e]
(13) KP × P	N–Q5	
(14) N × N	N × P	
(15) N–B5	N × N	
(16) Q × Q	R × Q	
(17) N–K7 ch	K–R1	[f]
(18) P–R5	R–K1	
(19) B–B5	P × RP	
(20) B–B4	P × P	[g]
(21) B × P	N–K5	
(22) B × R	B × P	
(23) R–QN1	B–B6 ch	[h]
(24) K–B1	B–KN5	
(25) B × KRP	B × B	
(26) R × B	N–N6 ch	[j]
(27) K–N2	N × R	
(28) R × P	P–R4	
(29) R–N3	B–N2	
(30) R–KR3	N–N6	
(31) K–B3	R–R3	
(32) K × P	N–K7 ch	
(33) K–B5	N–B6	
(34) P–R3	N–R5	
(35) B–K3	Resigns	[k]

B

HIERARCHICAL CONSIDERATIONS RELEVANT TO WHITE'S
PLAY

Level   *Remarks*

Note [a]:

N + 2   Play on the Centre, in Files,
         diagonals if possible.

N + 1   Select play in Centre, with pros-
         pects on diagonals.

N   Now find a move to strengthen
         Centre. This tactical move is
         N—B3.

Note [b]:

N + 2   Black's move has introduced new
         positional feature of an open
         diagonal for the Q-Bishop, and
         thus contests White's Q-diagonal.

N + 1   Contest the diagonal of Black's
         Q B, or play for control of diagonal
         by White's Q B, or play on open
         Q-file.

N   Test the (N + 1)-ideas via specific
         moves. Notice that Q–Q2 is
         strong since it achieves two goals,
         but White selects a move which
         contests the diagonal of Black's
         Q B.

Note [c]:

N + 2   Black's reply co-operates with
         White's strategy of a K-side
         attack by Pawn advances.

N + 1   Continue with the K-side attack
         if there are no tactical chances for
         Black.

N   Select the next Pawn move, P–N5.

Level   Remarks

Note [d]:

N + 2   Black's reply has suggested a counter-attack in the Centre (by P–Q4 or P–K4). Hence modify the (N + 1)-play to combine K-side attack with Centre control.

N + 1   Search for N-level (tactical) K-side moves and Centre control.

N   Find the Pawn move, P–B4.

Note [e]:

N + 2   Black's reply was consistent with his (N + 2)-strategy of Centre counter-attack, but he offers a Pawn. Go to N-level to check possible replies if accepting the Pawn.

N   P × P gains a Pawn, which is

N + 1   good at N + 2, but it opens the K-file and reduces the K-side attack, but it also threatens

N   P × P winning the Knight and so forces a Knight move in reply. Can the Black N on QB3 move to the K-side to help against White's

N + 1   attack, by N–K7? Test this and counter Black's P–Q5. Hence

N   choose Pawn exchange.

Note [f]:

N + 2   Material even, continue with (N + 1) positional K-side attack.

N + 1   Try to open a file by Pawn advance.

N   Select the Pawn move to R5.

Level	Remarks

Note [g]:

N + 2   Black has countered by an attack on the open K-file but has failed to contest the effect of White's KB. Are there possibilities of winning material via the K-side pressure?

N   B × P wins material but it also
N + 1   reduces the attack on the K-file.

N   Select B × P because material win is a good (N + 2) gain.

Note [h]:

N + 1   White's K must move to a square which is not on a diagonal accessible to Black's QB on his next move.

N   There is only one such square.

Note [j]:

N + 2   White must get out of check.

N   Select a white square for safety.

Level   Remarks

Note [k]:

N	The threat of P–N6 is very strong.
N + 2	It would be the fulfilment of White's strategy which began in the early part of the game. The
N + 1	helplessness of Black's position must lead to further material loss for him and eventual Mate.

I hope this illustrates to the reader that the ability to play chess need not be associated with any other specialized talent, such as that for prodigious calculation (or for mathematics generally). Indeed there seems to be no special corner of the talent market which dominates the chess scene. Some of the world champions have been far removed from the arena of science or technology. So we should not suppose that there is any special danger in developing, for example, computers which play chess – unless we can make them 'think' in the way that human beings do. At the moment this is not the case; certainly there are some good computer players around at the moment, but they are mostly designed on the principle that 'looking ahead' (which means tactical moves at the N-level) as far as possible is the only way to play the game. Chess Masters do not in fact work on that principle – because the human mind uses this hierarchical approach to chess, and to most of its other problems as well, and is not very good at looking ahead along a long line of tactical moves at one fixed hierarchical level.

This is why the problem of understanding how Masters play chess is highly relevant to the general problem of 'decision-making'. How do we make our decisions – in all the complicated situations in which we regularly find ourselves? If in fact we can understand the specific hierarchical structures which characterize such situations then we shall be a long way down the road to making decision-making into a rational (scientific) discipline.

Hard and soft science

By the early part of the fourteenth century intellectual satisfaction with the ancient theories of physics had waned considerably. The Aristotelian views on the laws of dynamics were not tenable to minds which felt a growing urge to understand the evidence of their eyes, in a commonsense way rather than in the context of some ontological theory. What later was to become one of the great successes of the eighteenth and nineteenth centuries, namely the theory of the motion of bodies under forces, was at that time still struggling to find its feet and it was not helped by the state of affairs in which high-level concepts about cause and effect dominated the thinking. The problem was that it was very difficult to express most of the Aristotelian ideas in mathematical terms. The idea of the 'first cause' of the motion, or of the 'second cause', of the 'place' towards which a body was assumed 'naturally' to move, was not obviously expressible in terms of the mathematics of the day – even though that mathematics was very simple compared with what it was later to become. When an idea was expressed mathematically, such as the idea that the force which caused the 'unnatural' motion was proportional to the velocity acquired, then it turned out to be wrong (in modern terms), even though it was inspired only by the need to be compatible with higher principles – such as the assertion that 'nature abhors a vacuum'.

Because of this difficulty about expressing the facts of motion, and any consequent theory of motion, in mathematical terms it is fair to describe the whole scene as a *soft science*. But in the fourteenth century some significant steps were taken in the direction of making the science harder. These were associated with a school of study in Merton College, Oxford, around the years 1350–70 and due largely to Thomas Bradwardine, William Heytesbury and Richard Swineshead. An interesting account of those times will be found in the history by A. C. Crombie [6] or in that by O. Pedersen and M. Pihl.[7]

What came out of the Merton School was the idea that we must make a clear distinction between (what we now call) *kinematics* and *kinetics*. The former is a *description* of the motion only, quite independent of the idea of its 'cause' (the *forces* which apparently cause it), whilst the latter combines this description with ideas of forces. Indeed William of Occam (he of Occam's Razor repute) went so far as to assert that kinetics was irrelevant and that only kinematics was needed to describe and to explain the motion of any body; he was effectively saying that the description was also the explanation. This role for kinematics was later to be emphasized in the context of gravitation by Einstein, in the twentieth century.

But kinematics becomes a question of geometry and of observation, of measurement of position (space locations in Euclidean geometry) and of time. That meant then, and right through to this century, 'measurement' of feet and inches and of seconds and minutes, and capturing the description (the kinematics) in these terms (which are essentially mathematical) was to make the whole thing into a *hard science*.

The central problem of Aristotelian physics being 'soft' seems to me to be illustrated very neatly by the sort of thing which had been said for a very long time and with wide acceptance about the motion of bodies. It had long been understood that 'motion is the realization of a body's potential' – a sentence with high-level words in it, not a sentence which was to be very useful to those engaged in building the hard science of dynamics. But what was it like to be around in the fourteenth century and to hear that sentence and to feel that it made a lot of sense, that it was even profound? Well, I suggest that you need only be in this century and that you move yourself into a modern soft science and make the statement that 'education is the realization of a body's (person's) potential'. That makes modern sense? It is just as soft as medieval dynamics – so where is the modern Merton School which will demand to know where the kinematics for such a science is to be found? Where is the geometry which will allow us to describe that thing 'education', where are the measurements and the observations which will make this statement of Social Science into a hard statement? Where are the mathematics?

Ah, you may well say, that is an illusion because it is clear to many social scientists that 'mathematics cannot be used to describe such a thing as education – *in principle*'. Just the very thing they were saying in medieval times! In those days people felt that 'motion' was a very complex thing; it had a great deal of poetry about it; the body in motion had a 'destiny' to fulfil, a 'goal' to reach, something which exhibited the 'great plan' of the divine creator. By representing it in terms of puny mathematical symbols, with numbers and calculations, all that subtle essence would be ignored or destroyed. Motion was a becoming, a growing and a blossoming, a realization of potential. Well yes, all these things are true, but that does not alter the fact that they are soft statements, nor does it alter the possibility that there can be a hard science waiting to come on the stage. When that happens the poetry will return – even though it will have to be written in the language of mathematics. After all, that is what happened with medieval dynamics, and the poetry of motion came to be expressed in the elegant and beautiful theorems of Newton, Hamilton, Lagrange and Einstein.

It took another two hundred years before the full import of what the Merton School had said became an actuality and produced a genuine hard science of dynamics. That happened with the work of Galileo[8] when he made the simple observations of rolling a ball down an inclined plane and measuring its position and time. What he had done was to introduce (perhaps unconsciously) the idea that 'measurement' is equivalent to the *identification* of *set-membership*. In other words he used a set of things he called 'points' or 'positions' and another set of things he called 'moments of time', and his 'measurements' consisted of ticking these members off, one by one, as the ball proceeded on its way. That is what made dynamics into a hard science – the idea of identifying *sets* of things as the basis of what we now call *data*. It is also ironic that in this we can see the hand of the master, Aristotle, who had stressed the notion of set and made much use of it in his various works – not least being that use he made of it in his *Politics*. If the early Schoolmen and medievalists had grabbed hold of *that* concept and applied it to correct Aristotelian physics then we might already be more than two hundred years ahead in our science. Not just in our technology – which is all about the dynamics of inanimate physical things – but in our social science (which is still in the soft stage), in our economic and political theory, in our psychology and medicine.

But the point of this sketchy historical study is to relate this notion of set-membership to our remarks about hierarchies (of sets of data), and this is now feasible because, quite independently of the practising scientists, mathematicians and logicians have provided us with intellectual tools for the study of 'sets' and incidentally warned us of some of the pitfalls. Much of this recent study remains in the classroom among logicians and philosophers and is perhaps as incestuous as any other academic field, but the work of Bertrand Russell in this context seems to ring out with a message which we can no longer ignore and which is telling us something vital for the scientific methodology we need now and shall need in the twenty-first century. He discovered the subtle logical and qualitative difference between a set of elements and the set you make out of collections of its elements (out of subsets of the first set). This is exactly the difference we have pointed up in our discussion of hierarchical levels N, N + 1, N + 2 etc., so far. It also emphasizes yet again the Aristotelian point that 'the set is greater than the sum of its parts' in the sense that the set is qualitatively different from (and qualitatively superior to) the mere collection of its members. By forming those members into a new whole (which is the *set* per se) we obtain something superior to the members themselves. A simple example illustrates the formalities of what we are talking about.

Take a set, called X, and let it have the following members:
{John, James, Sarah, Susan, Clarissa, Freddy}
This set can be at one of our hierarchical levels, say the N-level. Now a possible set of subsets of X will be the set X′ with members:

{{John, James}, {James, Sarah, Freddy}, {John, Susan}, {Clarissa}}

Here we notice that the members of X′ are *sets* of members of X; we notice too (though this is irrelevant at this point) that X′ is a cover of the set X. This set X′ we would put at the (N + 1)-level in the hierarchy. If the set X′ contained, as members, *all* the subsets of X (there would be $2^6 = 64$ such subsets) then we call it the *power set* of X and would write it as P(X). Obviously P(X) is also a cover of X, but it seems to rather overdo it since we can cover X with far fewer members than are contained in P(X). It is clear too that we can go on like this and form a set of sets of sets of elements of X; if we keep forming the power set we can neatly represent the possibilities by writing down the sequence of sets as X, P(X), $P^2(X)$, $P^3(X)$, $P^4(X)$, ... etc. where, for instance, $P^2(X)$ means the power set of the power set of X.

Now why did Russell warn us about dangers inherent in not keeping these kinds of sets distinct in our minds (and in our data)? It was because he discovered, in the course of writing his famous *Principia Mathematica* with his colleague Alfred North Whitehead, that there are serious logical paradoxes waiting to ensnare us if we confuse these levels. Russell referred to the levels as 'logical types' and his arguments about them are usually called Russell's Theory of Types. In particular he found that confusion about the levels could lead us to ask questions of the data which cannot be answered, which lead to logical contradictions. That is to say, if we regard members of P(X) as in the same logical category as members of X, and then ask some question of those members of P(X) and of X (treating them all alike), we can get into a paradoxical situation. To illustrate his meaning Russell introduced his famous 'Barber Paradox', as follows.

In a certain town all the men are clean-shaven and the male barber shaves all those men who do not shave themselves. Furthermore any man who shaves himself is not shaved by the barber. The question we would like answered is 'Does the barber shave himself?'

It seems simple enough: the barber is a man; he either shaves himself or he doesn't; so how can we answer the question?

Let us assume the answer to be Yes, the barber shaves himself. Then he is a man who shaves himself, so he cannot be shaved by the barber, so he does not shave himself. We must have been wrong, the answer must be No. But then, in that case, the barber does not shave himself, so he is a man who is not shaved by the barber, and so he must shave himself; so the answer is Yes; again we logically deduce a contradiction.

This is exactly the problem of confusing the levels, for we can represent the situation by a simple relation between the barber (call him B) and the possible set of men in the town (call them M_1, M_2, M_3, M_4, M_5, M_6, M_7, M_8) – the number is not important. In the table below, which represents the relationship 'is shaved by the barber', we notice that the 1's indicate those men who are shaved by B and the 0's those who are not. The question cannot be answered because we cannot decide whether or not to put a 1 or a 0 underneath the B in the top row – and after all the barber ought to be there somewhere, being a man who has to be shaved.

	M_1	M_2	M_3	M_4	M_5	M_6	M_7	M_8	B
B	1	0	0	1	1	0	0	1	?

The point is that the man who is the barber is only a barber (is only defined properly) in so far as he shaves people. The word 'barber' (or the symbol 'B') therefore really represents the subset of men, namely, the *set* $\{M_1, M_4, M_5, M_8\}$; so 'barber' is at the level $N + 1$. When we try to put the B in among the M's we fail because we are trying to regard an $(N + 1)$-thing as an N-thing, and this leads to the paradox. So Russell's Theory of Types corresponds closely to our hierarchical levels in the following obvious way:

type:	X	P(X)	$P^2(X)$	$P^3(X)$	etc.
level:	N	$N+1$	$N+2$	$N+3$	etc.

The lesson to be drawn from this is that we must expect our data to be hierarchically organized, in the way we have already indicated, before we can expect to use it as the basis for any hard science. We have therefore effectively defined hard science as that which is based on the use of *hard data* (data which is associated with set-membership – it being understood that the sets must be hierarchically organized via our notion of cover-sets). This is the essence of scientific data – as it has been collected and used during the past two hundred years in the physical sciences. The significance of this analysis is greater, now that we are moving into an era when Social Science needs a paradigm in the Kuhn sense.[9] The fact that physical science has had this property for such a long period and that the property has been largely overlooked is no doubt due to the fact that, in something like physics, this notion of set-membership has been relegated to the laboratory where scientific instruments have performed the chore of sorting things out into their appropriate sets. The galvanometer is a device for sorting out electrical charges from non-electrical charges. Indeed in the early days of the Cavendish laboratory at Cambridge, when Cavendish was the professor of experimental physics, this sorting-out was often done by the scientists themselves. Cavendish had a reputation for being able to give a pretty accurate estimate of the quantity of electrical charge present in a charged device merely by grasping the terminals in his hands. The quiver he felt was a measure of the charge; presumably the development of the galvanometer as a testing instrument saved a lot of lives.

But now the reader must be warned about the *sensations* associated with becoming 'scientific' in this *hard* sense, for most of us spend most of our time being quite definitely *soft* in the head (in this soft-science sense, of course). Making 'hard science' out of the motion of bodies (the science of mechanics – flourishing as it did during the eighteenth and nineteenth centuries) certainly meant that attitudes and values had to change drastically. In one sense even the simple thing of a cannon-ball flying through the air had to lose a lot of what men would have once called 'its poetry' (the Aristotelian 'purpose'), and the *sense* of that poetry would have been seated somewhere near the solar plexus region of your average man. His awareness of the phenomenon would have been via what Jung would call his 'feeling-Self' – and he judged it (valued it) by having a 'feeling' about its 'effects'.

This was bound to be 'soft' because, in order to become 'hard', it would have to be judged and valued by the 'thinking-Self' (whose currency is always *concepts*, not *emotions*). So when it becomes 'hard' it somehow moves up into the head – and becomes emotion-less, and the observer loses contact with that previous sensation (which we are calling 'soft'). In that sense, therefore, something is lost – and we naturally feel nostalgia for that loss, and we ourselves have become something 'less' than we were before (in that context). But at the same time we have become more than we were before, for now the 'hardness' has given us a new world with its own 'poetry' – in this thesis we are claiming that it is an essence of *structure* (and an associated multi-dimensionality which we shall progressively explain throughout this book) which gives us the new scope for exercising our 'feeling-Selves'. This is what the world of science is really all about – and if it seems to fail us at any time it can only be because the science has not gone deeply enough into the phenomena, has not found that structure which is adequate for our emotional needs. Certainly it seems to me that we are very much in that position today, when 'science' is largely bankrupt of ideas for coping with *us* (and our social Selves) – for mechanistic ideas about 'particles' seem to relegate us to a desert (which no one dare water?).

Going to the Moon

What was the really profound significance of the struggle between the geocentric and the heliocentric theories of the planetary system? Why did the view which originated with Copernicus during the Middle Ages finally triumph over the views which were associated with

Aristotle and the Ptolemaic system? Why was Galileo challenged by the established church and eventually forced to renounce his belief in the heliocentric theory? Surely it could not have been simply a matter of the authority of the established church? Why did it matter so much to the 'scientific' cardinals of Galileo's day whether or not the earth was at the centre of the planetary system? Did that belief belittle God or did it belittle Man? Somehow the question must have dug deeper into the human psyche, must have made the cardinals feel insecure in their dogma. Something about it must have frightened the daylights out of them, must have threatened to bring the whole system crashing down about their ears. Well, what was it, this terrifying threat?

The heliocentric theory certainly challenged that philosophy which placed Man at the centre of things, Man as the divine creation, Man as the centre of attraction, as the *highest level* of divine achievement. Writers and historians seem to agree on this, that the new theory required the removal of Man from the centre; but what was the significance of removing Man to the periphery? Surely it was of a *hierarchical* significance and dependent on the prison which the scientific cardinals had built for themselves and for all mankind, a prison which confined the imagination and imposed a hierarchical ceiling on Man's estate. Looking through Galileo's telescope at the Moon took Man away from this prison and let his spirit soar upwards. For if Man is to be at the centre of things, if all is to be subordinated to Him as the prime mover and object of divine affection, if the place that He inhabits is to be the centre of all planetary action, then all these ideas are only compatible with the other idea that Man himself is trapped at some hierarchical level, say the N-level. For if He is trapped at this level, then He is quite incapable of being 'outside of himself' and thereby seeing himself only as some particular N-level set of entities. For if Man is to be outside himself and see himself in a new context he can only achieve this by moving up the hierarchy to (N + 1) and by looking down on the previous N-level. In that new condition he acquires a new freedom of action and a new freedom of thought, because he acquires the freedom to rearrange everything at the N-level and to see it with a new cover. This cover, of course, refers to a set of things at the (N + 1)-level. Hence he cannot rearrange the N-level without moving himself into the (N + 1)-level. Perhaps this movement should not be completely described as an intellectual one, perhaps it would be better to allow the use of the word 'spiritual' to describe this movement. But maybe this is only evading the question? Perhaps the word 'spiritual' is itself an indication of the ability to move into higher hierarchical levels and so look down onto a situation at lower levels which can now be rearranged and seen in a different light. Is not this the essence of all profound scientific discovery?

The same kind of idea can be traced to those developments at the beginning of the twentieth century in which the old classical Newtonian mechanics was replaced, to a large extent, by the novel theories of relativity propounded by Albert Einstein. In postulating that the world should be viewed as a four-dimensional continuum, Einstein had moved into a higher hierarchical level than was necessary to accept and to appreciate the hitherto classical theories of existence in a three-dimensional space. In that three-dimensional space, in terms of the classical Newtonian mechanics, dynamics consisted of an experience in a three-dimensional space and that experience was manifest through a certain concept of time. If we regard the expression of that dynamics as associated with an N-level set of concepts then, coming to the Einstein theory, we must express the new concept as existing at an $(N + 1)$-level. For at that $(N + 1)$-level it is now possible to rearrange the concepts of space and time into new conglomera known as the space-time continuum. Thus the development of scientific thought, via these Einstein ideas, seems to be to express this searching for and achieving of a higher hierarchical view of our sense-data.

Bringing ourselves up to date, we must admit that the space research programme which has resulted in sending human beings to the surface of the Moon has for its proper justification only this idea of moving men, with their intellects, up the hierarchical ladder to a new level. For primarily we shall not achieve a greater intellectual emancipation without moving up this hierarchy. In metaphorical terms, it means we must mentally travel to the Moon. If we cannot make that effort to travel to the Moon mentally, or its equivalent – of moving up the mental hierarchy into new concepts which act as a cover of lower concepts, then we must be carried there forcibly in our boots. We must in fact be made to stand on the surface of the Moon in order to look back and see the Earth and ourselves in a new hierarchical concept.

Hence, although it is easy to say that the millions of dollars which have been spent to send us to the Moon could have been more profitably used to alleviate suffering and misery on this Earth, this can only be true in a limited hierarchical sense. For in order to alleviate suffering and misery on the surface of the Earth it is necessary to have a view of Mankind and his role on that surface which is a higher hierarchical view than we have had heretofore. Only with a new hierarchical view of Man and his problems will we be able to see a way through to a solution to these problems. We shall never solve N-level social problems if we can live only at the N-level. It is essential to move to the $(N + 1)$-level – and possibly to the $(N + 20)$-level. The progress which has resulted in the technological victory necessary to send men physically to the Moon can be seen in this light, that by this process we achieve a new hierarchical position from which to solve Man's problems on the surface of the Earth.

Is this not the real explanation why, in our present time, the work of science-fiction writers has become a leading fictional and cultural enjoyment for the intellectuals? That intriguing trilogy by Asimov[10] illustrates very forcibly this need for man to move out of his present confines – that is to say, out of the trap of being at some fixed N-level in the hierarchy – into a new hierarchical set of levels which allows him to look back on his present condition and to see it 'in perspective'. The science-fiction writer is therefore fulfilling the need which is also associated with an increasing tempo of events, for projecting ourselves forward into a new (and at this stage fictional) age of hierarchical levels which we have not yet built into our actual way of life. In this respect the science-fiction writer, at his best, is an important prophet for the rest of mankind. His prophecies are not so much the prophecies of technological achievement but the prophecies of new hierarchical levels of concepts. In these hierarchical levels of concepts, which we have yet to explore, are we sure to find our salvation?

2. Laughter and Tears

In the Soviet Union the game of chess is one of the most popular pastimes and it is inevitable that it should therefore be taken very seriously by a large number of people. Equally it is not surprising that it can become a subject for a satirical joke, and in their humorous novels the Russian writers Ilf and Petroff have depicted a character called Ostap Bender. This character is an idiot at playing chess and yet he spends his time giving elaborate lectures on the game. Naturally these exploits of Bender in establishment chess circles serve as a source of amusement and comic repartee. In a very serious book on chess the Russian Grandmaster Alexander Kotov quotes from one of Bender's imaginary lectures where, in answer to a question about the possibility of women learning to play chess, he replies, 'The blonde plays well and the brunette plays badly, and no amount of lectures will change this state of affairs.'

I find this amusing, although not uproarious, but sufficiently so to ask why. Although it might be dangerous ground to explore, nevertheless I intend to argue that the idea of its being amusing, the fact that it generates laughter, is closely tied up with the idea of hierarchical levels of consciousness – such as we have already discussed in the first chapter.

In the first instance the anecdote is about certain people (particularly chess-players) who are given the names Blonde and Brunette, together with a whole set of unspecified concepts which are involved in an understanding of the word 'play'. All these concepts we can regard as being at some hierarchical level, say the N-level. Now the context of this witticism is an apparently serious lecture in chess given by Ostap Bender. Therefore we shall suppose that the entities which are named are various things associated with the game of chess. Also there are to be entities such as the listeners to the lecture. Now suddenly the lec-turer begins by referring to these words, namely 'Blonde' and 'Brunette', which can be properly identified as elements in this N-level set and elements which constitute a proper subject for this serious lecture. This is because it is just conceivable that the human beings commonly referred to as Blonde and Brunette are possible candidates as opponents in the game of chess – these might be unusual ways of referring to them but they are just about acceptable. The use of the word 'plays' immediately lifts us into the next hierarchical level $(N + 1)$, because it refers to all possible kinds of games. Hence this reference to how the Blonde and the Brunette play allows us to shift our attention sideways, as it were, to cover a new set of possibilities, things which do not refer to chess but which can still properly be contained under the word 'plays'. Hence, without labouring the point too much, the witticism immediately requires us to be aware of an $(N + 1)$-level state of events. From that position we are immediately able to contemplate new relationships on the N-level set – either by rearranging the elements that we already had in mind, or by extending the elements so as to find new relationships between them. I am claiming that it is this elevation *suddenly* to the $(N + 1)$-level which generates a release of laughter in us – rather than a sudden realization of what it is that the Blonde and the Brunette 'might be good at playing'. It is as if moving up the hierarchical separate levels in a sudden leap releases a certain kind of energy (without searching for another word at the moment) which is itself manifest as *laughter*. Although we have not yet learnt how to identify the proposed multidimensional structure at any particular hierarchical level it is worth pointing out that, if we assume that such a structure may be shown to exist, then the number of dimensions (which express the different complications which can be experienced at that level) will normally drop as we move up the hierarchy – because of the simplifying action of finding cover-sets. In this sense a sudden insistence on moving up the hierarchy must act as a sudden release from a certain complicated and higher-dimensional space to another slightly less complicated and lower-dimensional space. It may well be, therefore, that the energy which is released is itself some strange psychological energy which is built into the multidimensional structures which are to be found at the different levels. However, at this stage, we do not wish to pursue the details of the actual multidimensional structure. We need only examine, as a plausible idea, the production of laughter, the appreciation of humour, the sense of what is funny, in various particular instances. In order to have some point of reference we are including a scheme which is shown below to remind us of the outline of the theory. It will be seen then that laughter is emitted when one moves up the hierarchy and conversely we shall designate what is absorbed (as opposed to being emitted) by the word *tears*. Thus laughter is a sign of a sudden movement up the hierarchical level whilst tears are a sign of a sudden fall down the hierarchy. Movement *upwards* corresponds to a widening of our conceptual horizons, a process of aggregation, whilst movement *downwards* must consist of a shrinking of those horizons. Laughter is a sign of tolerance, 'to know all is to forgive all', whilst tears are a sign of shrinking and contraction, of being unable 'to see a way out', of intolerance of oneself (of one's plight).

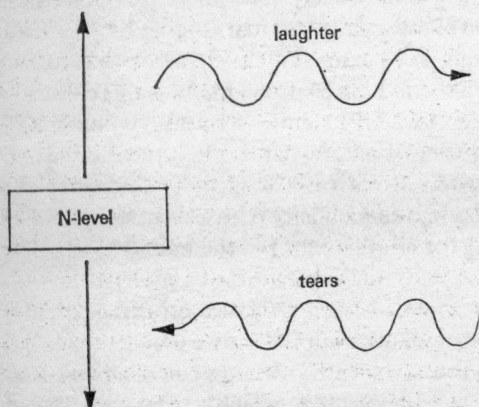

Fig. 3 Hierarchical jumps

This idea that, by moving up the hierarchical level, one suddenly becomes aware of new relationships at the first level can be illustrated from the work of a variety of writers of humour. It can be, for example, the kind of description which involves a complete inversion of accepted values of what comes first and what comes second. And it can be illustrated either by a short but subtle rearrangement of meaning in a simple sentence or by a work of extended illustration of it in a complete novel. Here we might well note the words of William Hazlitt when he wrote: 'Man is the only animal that laughs and weeps; or who is the only animal that is struck with the difference between what things are and what they might be.' The American humorist Mark Twain wrote that the 'secret source of humour is not joy but sorrow; there is no humour in heaven'. If to be in heaven is to know the ultimate state of awareness, almost the infinite hierarchical level, then there can presumably be no more levels to reach? It would be sad but true that there could then be no humour.

Here it is interesting to ask oneself why one says it would be 'sad' that such a state would prevail. This can surely only reflect the deep and abiding need in every man, including the author, to believe that there are yet higher hierarchical levels available for him to reach. His sense of humour is his only anchor in that belief, his firm conviction that it is possible to keep moving and every time that he needs to inquire he needs also to find some trigger which will set off his sense of humour. Hence he needs to be able to find something higher in order to accept the notion of a 'joke', and ultimately he must be able to find something funny even in the idea of all possible jokes. If he is to reach an ultimate hierarchical level he would presumably have also reached the end of all possible jokes!

Perhaps too it is appropriate here to append the opinion of Christian Morgenstern on the matter: 'Humour is the contemplation of the finite from the point of view of the infinite.' If we replace his use of the word 'infinite' by $N + k$ and if we replace his use of the word 'finite' by N (and just let k be a reasonably large number) then is that opinion not the same as this hierarchical theory of laughter?

Joseph Heller

There is that haunting humour to be found in Joseph Heller's *Catch-22*.[11] Here the theme runs through the length of the work but it is no better illustrated than in the following extract:

Yossarian looked at him soberly and tried another approach.
'Is Orr crazy?'
'He sure is,' Doc Daneeka said.
'Can you ground him?'
'I sure can. But first he has to ask me to. That's part of the rule.'
'Then why doesn't he ask you to?'
'Because he's crazy,' Doc Daneeka said. 'He has to be crazy to keep

flying combat missions after all the close calls he's had. Sure, I can ground Orr. But first he has to ask me to.'

'That's all he has to do to be grounded?'

'That's all. Let him ask me.'

'And then you can ground him?' Yossarian asked.

'No. Then I can't ground him.'

'You mean there's a catch?'

'Sure there's a catch,' Doc Daneeka replied. 'Catch-22. Anyone who wants to get out of combat duty isn't really crazy.'

There was only one catch and that was Catch-22, which specified that a concern for one's own safety in the face of dangers that were real and im-mediate was the process of a rational mind. Orr was crazy and could be grounded. All he had to do was ask; and as soon as he did, he would no longer be crazy and would have to fly more missions. Orr would be crazy to fly more missions and sane if he didn't but if he was sane he had to fly them. If he flew them he was crazy and didn't have to; but if he didn't want to he was sane and had to. Yossarian was moved very deeply by the absolute simplicity of this clause of Catch-22 and let out a respectful whistle.

'That's some catch, that Catch-22,' he observed.

'It's the best there is,' Doc Daneeka agreed.

Here Yossarian can be regarded as one of a number of individuals at the N-level. At the (N + 1)-level there is a set containing at least the following five words:

{Insane, Sane, Flying Missions, Grounded, Fit-for-duty}.

At the (N + 2)-level there is for practical purposes a single name which Yossarian thinks of as the Doctor or simply Doc. Now this Doc has the responsibility for ensuring that the (N + 1)-level set has a suitable cover for the N-set. That is to say, he can decide whether a man at the N-level is properly a member of the thing called Fit-for-duty at the (N + 1)-level and this has implications for whether that man is also to be found in the element 'Flying Missions' or 'Grounded'.

Level	Set
N + 2	{Doc}
N + 1	{Insane, Sane, Flying Missions, Grounded, Fit-for-duty}
N	{Yossarian, Orr, etc.}

Now, Yossarian is convinced that he is covered by the two words 'Insane' and 'Flying Missions'. This makes him think that he cannot be also covered by the word 'Fit-for-duty'. Hence he goes to the Doc and asks to be grounded on the grounds that he must be insane to consistently expose himself to danger by taking part in the flying missions. Now the Doc immediately reshuffles the cover set at $(N + 1)$ by saying that this request by itself is enough to demonstrate that Yossarian is sane – because of this sudden flash of rational thinking that he has experienced. Hence the Doc takes him from the Insane and puts him in the Sane subset. This ensures that he is fit-for-duty and therefore eligible to take part in the flying missions. For his part Yossarian feels that if he is in flying missions he must be insane and therefore goes round in this vicious circle which is referred to as Catch-22. When reading the passage and identifying oneself with Yossarian one can only feel the anguish which comes from the sense of being trapped. This trap can be nothing more than the trap of being inside the N-level and unable to get out. By appealing to the $(N + 2)$-level for a rearrangement of the cover-set and being rejected, Yossarian is being trapped. This sense of trapping holds him at the N-level. It does not generate tears in him because it does not force him down to the $(N - 1)$-level, but it generates a kind of mid-tears-cum-laughter combination of anguish and frustration. The reader is allowed to move up the hierarchy to $(N + 2)$, and even to $(N + 3)$ sensing the authority that lies above the Doc, and in so doing is encouraged to feel that release of energy which is a sense of humour and gives rise to laughter. This is a not uncommon situation in which the reader hardly knows whether to laugh or to cry. If he identified himself with the hero, in this case Yossarian, then he shares that anguish we have mentioned and is on the verge of tears. But if he suddenly is aware of himself as a reader, he is allowed the freedom to move up to $(N + 3)$ or even higher. In that condition he may experience the need to laugh. It is this intriguing quality of the episode which makes it so valuable an experience.

Jaroslav Hasek

Another prime example, almost the supreme one of the reader's dilemma, is to be found in that great fictional work by Jaroslav Hasek, *The Good Soldier Schweik*.[12] We give as an example the following extract:

'SCHWEIK AS A MALINGERER'

At this momentous epoch the great concern of the military doctors was to drive the devil of sabotage out of the malingerers and persons suspected of being malingerers, such as consumptives, sufferers from rheumatism, rupture, kidney disease, diabetes, inflammation of the lungs, and other disorders.

The torments to which malingerers were subjected had been reduced to a system, and the degrees of torment were as follows:

1. Absolute diet – a cup of tea morning and evening for three days, accompanied by a dose of aspirin to produce sweating, irrespective of what the patient complained of.

2. To prevent them from supposing that the army was all beer and skittles, they were given ample doses of quinine in powder.

3. Rinsing of the stomach twice daily with a litre of warm water.

4. The use of the clyster with soapy water and glycerine.

5. Swathing in sheets soaked with cold water.

There were dauntless persons who went through all five degrees of torment and had themselves removed in a simple coffin to the military cemetery. There were, however, others who were faint-hearted and who, when they reached the clyster stage, announced that they were quite well and that their only desire was to proceed to the trenches with the next draft.

On reaching the military prison, Schweik was placed in the hut used as an infirmary which contained several of these faint-hearted malingerers.

'I can't stand it any longer,' said his bed-neighbour, who had been brought in from the surgery where his stomach had been rinsed for the second time.

This man was pretending to be shortsighted.

'I'm going to join my regiment,' decided the other malingerer on Schweik's left, who had just had a taste of the clyster, after pretending to be as deaf as a post.

On the bed by the door a consumptive was dying, wrapped up in a sheet soaked in cold water.

'That's the third this week,' remarked Schweik's right-hand neighbour. 'And what's wrong with you?'

'I've got rheumatism,' replied Schweik, whereupon there was hearty laughter from all those round about him. Even the dying consumptive, who was pretending to have tuberculosis, laughed.

'It's no good coming here with rheumatism,' said a stout man to Schweik in solemn tones, 'rheumatism here stands about as much chance as corns. I'm anaemic, half my stomach's missing and I've lost five ribs, but nobody believes me. Why, we actually had a deaf and dumb man here, and every half hour they wrapped him up in sheets soaked in cold water, and every day they gave him a taste of the clyster and pumped his stomach out. Just when all the ambulance men thought he'd done the trick and would get away with it, the doctors prescribed some medicine for him. That fairly doubled him up, and then he gave in. "No," he says, "I can't go on with this deaf and dumb business, my speech and hearing have been restored to me." The sick chaps all told him not to do for himself like that, but he said no, he could hear and talk just like the others. And when the doctor came in the morning, he reported himself accordingly.'

This is almost too painful to describe, but it is pretty clear that it is in the same range as the previous quotation from *Catch-22*. Only by lifting oneself into the (N + 2)-level out of the need to be a typical inmate of this extraordinary medical establishment, can one even begin to bear the (N − 1) details of the treatment of the so-called malingerers. In fact it is only possible, I would suggest, to move up to (N + 2) by glossing over the words, which describe the (N − 1) punishments, as if they are faint symbols of a reality which is happily obscured from our view.

Level	Set
(N + 2)	{Supra-plot-concepts}
(N + 1)	{Medical authorities}
N	{Tuberculosis, kidney-disease, ... good-health}
(N − 1)	{Schweik, other inmates} ↔ {punishments}

Schweik and the other inmates, together with the punishments, can be placed at (N − 1). Then at the N-level we have the medical diagnoses; these cover the people-cum-punishments at (N − 1). At the (N + 1)-level we can place those concepts (activated by the medical authorities) which amount to allotting the (N − 1)-words to the N-words. This process places Schweik in an appropriate cover-word and has as its guiding light the notion that all the words except 'good-health' are superfluous to the N-set. An inmate, such as Schweik, is challenging the (N + 1)-authorities by placing himself at that level and thereby pronouncing on the cover-set at the N-level. The authorities at (N + 1) counter this challenge by introducing additional things into the (N − 1)-set (these additional things are the punishments – the stomach-pump, etc.). They are effectively saying, 'You are not just the word "Schweik", you are in reality the compound word "Schweik-plus-stomach-pump".' They follow this with the question, 'Does this force you downwards from the (N + 1)-level and to that place at the (N − 1)-level (under "good-health") that we have allotted you?'

Reading this passage, without the agony of being at (N − 1), is only bearable by moving up to a set of concepts, at (N + 2), which are supra-plot. Such concepts act as a cover of 'all plots' and so are capable of coping with the particular fictional plot – indeed only thereby can this particular description be seen as fictional. This experience of jumping from (N − 1) to (N + 2) can release the energy of laughter – but the movement must be quick, otherwise one sinks down to (N − 1) and weeps.

Stephen Leacock

A less agonizing example of the hierarchical basis of humour, illustrating yet again the sideways movement at the N-level, made possible by moving to the (N + 1), is to be found in the classic short study by that profound Canadian humorist Stephen Leacock.[13] This is the study known as 'Boarding-House Geometry'.

'DEFINITIONS AND AXIOMS'

All boarding-houses are the same boarding-house.

Boarders in the same boarding-house and on the same flat are equal to one another.

A single room is that which has no parts and no magnitude.

The Landlady of a boarding-house is a parallelogram – that is, an oblong angular figure, which cannot be described, but which is equal to anything.

A wrangle is the disinclination of two boarders to each other that meet together but are not in the same line.

All the other rooms being taken, a single room is said to be a double room.

'POSTULATES AND PROPOSITIONS'

A pie may be produced any number of times.

The Landlady can be reduced to her lowest terms by a series of propositions.

A bee line may be made from any boarding-house to any other boarding-house.

The clothes of a boarding-house bed, though produced ever so far both ways, will not meet.

Any two meals at a boarding-house are together less than two square meals.

If from the opposite ends of a boarding-house a line be drawn passing through all the rooms in turn, then the stovepipe which warms the boarders will lie within that line.

On the same bill and on the same side of it there should not be two charges for the same thing.

If there be two boarders in the same flat, and the amount of side of the one be equal to the amount of side of the other, each to each, and the wrangle between one boarder and the Landlady be equal to the wrangle between the Landlady and the other, then shall the weekly bills of the two boarders be equal also, each to each.

For if not, let one bill be the greater.

Then the other bill is less than it might have been – which is absurd.

This clearly requires us to move sideways across the N-level set of words which include experiences of a formalized mathematical education (in old-fashioned Euclidean geometry) and also in the apparently more homely experiences of trying to live in a boarding-house. There are of course overtones of these experiences which are dated and have that rather Edwardian feeling – which correspond to the time that Leacock was writing. It is clear that no elaborate analysis is needed for this whole article to be seen as an illustration of the hierarchical theme.

Level	*Set*
$(N + 1)$	{Concepts of relations, methods of proof}
N	{Relations between N-level elements (as in Euclid)}
$(N - 1)$	{Euclidean elements (line, square, etc.), . . . boarding-house elements (landlady, pie, etc.)}

If we allow that the details of life in a boarding-house and the details of elementary geometry are like an $(N - 1)$-level, then we have the interesting and amusing illustration here that not only do we need to move to the N-level in order to redistribute the $(N - 1)$-words but also we should move to the $(N + 1)$-level in order to accept the method of Leacock's 'proof' that the two bills shall be equal – each to each. This is because the method of proof, and the concept of a proof as found in Euclid, must be at least at the N-level, being as it is a study of some relation of things at the $(N - 1)$-level. Now we are asked to appreciate the sideways movement at the N-level of 'proof'. That proof is now applied to the charges which the landlady demands of the two boarders. To move into this conceptual position requires us to move up into $(N + 1)$, at least. Thus the comic nature of this proof certainly seems to require an immediate movement from $(N - 1)$ to $(N + 1)$.

Stephen Potter

One of the most fascinating, and peculiarly English, examples of modern humour is that to be found in the work of Stephen Potter. His *Gamesmanship*, *Lifemanship*, *Oneupmanship*, etc., are all first-class examples of hierarchical comic structure.[14]

Written after the last war, they constitute a study of what we call 'ploys' in an abstract game of living. This 'game' is a reflection on a certain kind of social smart set and is an illustration of forcing the interaction of the people down into a suitable hierarchical level at which all of life's irritations can be viewed with the detachment that one would normally reserve for viewing a game. This is equivalent to moving the reader up the hierarchy so that he may take a games-view of what he would normally take quite seriously. Thus the question, whether someone has hurt his feelings by an unfortunate remark, or whether he has been socially snubbed, or whether he feels that he has lost face in some social competition such as a silly game, is all reduced to the level of an abstract and inconsequential ploy. In this sense the works act as a kind of release from the pettiness of social life, and they do so by encouraging the reader to move up the hierarchy into a new position. We give only a short extract from Potter's work *Lifemanship*. It does not need a context setting and is probably self-explanatory.

'NOTE ON O.K.-WORDS'

My use of the word 'diathesis' reminds me that this is now on the O.K. list of Conversationmen. We hope to publish, monthly, a list of words which may be brought in at any point in the conversation and used with effect because no one quite understands what they mean, albeit these words have been in use for a sufficiently long time, at any rate by Highbrowmen, say ten years, for your audience to have seen them once or twice and already felt uneasy about them. We are glad to suggest two words for November:

Mystique

Classique

I have often been asked whether there is an accredited counter for use against O.K.-words. Mrs Johnstone made a note of the following conversation between myself and J. Compton, the educationist (Lifeman 364). Compton used to do splendidly with the word 'empathy' when it was O.K. in the twenties, but we are none of us as young as we were. He was trying a fairly up-to-date O.K.-word which has been on our list since October 1938: 'Catalyst'.

COMPTON: I think Foxgrove acts as a useful catalyst to the eccentricities of his chairman.
SELF: Catalyst?
COMPTON: Yes.
SELF: Yes. I suppose 'catalyst' isn't quite right.
COMPTON [*surprised*]: Not quite right?
SELF: Not quite what you mean. A catalyst is an agent of redistribution, literally.
COMPTON: Oh. Yes.
SELF: It is a re-alignment of the molecules rather than an alteration of their potential . . .
COMPTON: In a sense . . .

Compton knows, and I know that he knows, that I am as ignorant of physics or chemistry as he is; yet nothing he can say will alter the general impression that in the feverish pursuit of the O.K.-word he has misfired with a metaphor, ployed by his own gambit.

Level	Set
$(N + 2)$	{Reader of Potter's works}
$(N + 1)$	{Lifemen, Highbrowmen, Conversationmen, Scholars}
N	{O.K.-words (various), Compton, Self, . . .}

Not only do we here need to move from N to $(N + 1)$, in order to cover the particular conversation, but also we need to be a suitable 'Reader' at $(N + 2)$. At this $(N + 2)$-level the Reader must know what it is like to be (something of) a serious scholar, as well as a Conversationman. For only then can he appreciate all this talk about 'Mrs

Johnstone made a note . . .' (like a serious case-worker) and 'J. Compton, the educationist (Lifeman 364)' (all phoney mock-scholarly stuff – notice the J in J. Compton, greatly superior to John Compton or merely Compton).

Alexander Solzhenitsyn

Now what about the tears? What about the inverse of the laughter process, or other process which sends us down the hierarchical level, filling us with sorrow and sadness as we go? The real heartache in Alexander Solzhenitsyn's work *One day in the life of Ivan Denisovich*[15] lies in the problem of identifying oneself, as reader, with the 'hero' Shukhov. In order to do this properly one is compelled to move down the hierarchical levels and into the prison camp with our hero, into that life in the camp which becomes a daily routine of $(N - 1)$ or even $(N - 2)$ details. What is more, the whole aura of prison life in that camp is one in which the prisoners are kept at that low level. Thus Shukhov must spend all his energies and all his waking hours in the incessant struggle with the tedious and humiliating details of survival in the prison camp. Why are these humiliating? Because they reduce and maintain Shukhov at the lowest hierarchical level. How is he to secrete an extra piece of bread about his clothing so as to have something to eat later in the day? How is he to avoid the search by the camp guards after he has secreted a small broken penknife in his clothing – a penknife which will allow him later to mend other prisoners' boots and therefore to be rewarded by extra food? These are profoundly humiliating, because they are symptomatic of the fact that prisoners must not be allowed under any circumstances to move into a hierarchical position where they can look down on this life at the $(N - 2)$-level and see it in a detached way from the $(N - 1)$-level. For if one could do that it might well generate hope – a condition in which one could contemplate moving up the hierarchical level, thereby reducing the petty problems of life to such a small proportion that they disappear from the horizon altogether. This seems to me to be the essence of the cruelty inherent in the prison life. Taking away a man's 'freedom' is not really a matter of taking away his choice of work or his choice of friends or his choice of habitat. In its most profound sense, in its most wicked sense, it consists of taking away from a man the right and joy of living his life at whatever hierarchical level of awareness he may freely choose. All severe prison regimes seem to have this in common – that by a calculated programme of personal humiliation the prisoners are reduced to the lowest possible hierarchical level – and kept there, excluding hope and humour.

This must surely be the essence of the evil of slavery? It could always be argued, and it frequently was, that the American slave-masters in the last century were often kind to their slaves and consider-ate to their workers. But this meant being considerate about the $(N - 2)$ details of life. Do the slaves have a sufficiency of staple diet? Do the slaves have an adequate housing condition? Are the slaves being overworked? All these things could be put right by a liberal-minded slave-owner. But what could never be put right was the fact that the slaves, their slaves, were forced to live at the lowest hierarchical level. Because they were adult beings and because they had innate hierarchical ambitions they had to be tranquillized by the injection of suitable religious education, in itself a legitimate higher-level promise; in a land of Christianity, the well-intentioned slave-owner could inject into the black slaves a large concern for religious concepts – concepts which naturally referred to the $(N + 3)$- or $(N + 4)$-levels of human aspirations. This did not mean that the liberal-minded slave-owner was necessarily being in any way hypocritical. He prob-ably believed that the owners were superior people – which in fact they apparently were; but nevertheless this cannot disguise the inherent contradiction in the system – a contradiction which in modern times is related to civil strife. The black descendants of this day often reject what they contemptuously describe as 'White Liberals'. The aspira-tions of the underdog, such as the American Negro slave and his descendants, can only be aspirations for the ultimate human feeling of being able to move up and down a hierarchical level of awareness which he has freely chosen for himself and the limits of which can only be found in his own personal constitution. This must surely be the cry of all suppressed people and of all underdogs. The trap that they find themselves in is that trap which condemns them in a sub-merged hierarchical level. All the fake words are expressive of the same concept, words such as 'at the bottom of the ladder', 'know your place, man', 'the lower orders', 'getting uppity', etc.

But the North American continent provides us with the tragedy of both the black slave and the 'Red Indian'. It is no accident that there is no humour of any kind in Dee Brown's book, *Bury my Heart at Wounded Knee*.[16] This tale of the sad decline and fall of the American Indians before the advancing white settlers can only be a tale of sorrow and of tears. This is because it is a tale of the destruction of the hierarchical structure of the Indians' spirit. They had to abandon the hunting ground, and that meant that they had to abandon the *concept* of the hunting ground. That was the reality of their disaster. They had to move on to the reservation and to live by receiving handouts from white governments and military outposts. To accept this as his way of life, a way in which his independence of spirit and his ability and need to look after himself in a proper hierarchical context, of his own making, was ruthlessly destroyed – this had to be disastrous. Such a tale of weeping is hardly bearable. In the words of *The Times* review it is a book 'calculated to make the head pound, the heart ache and the blood boil'.

Our collective Selves

Tears which are the result of dropping suddenly down the hierarchy are also manifest through the many levels which go to make up our collective selves. There will be many such levels, with finer gradings, but an outline is provided by the following list.

Level	Collective Self
$N + 3$	The nation-self, or large-scale culture-self
$N + 2$	The region-self, or specific culture-self such as 'literature' or 'theatre'
$N + 1$	The workplace-self, or social-group-self
N	The family-self, or friends/loved-ones-self
$N - 1$	Self (the individual personality, the ego)

If we begin at the Self as the individual ego, the one we are supposed to know all about by first-hand experience, then the next level can be that at which we find a collection of such egos. This new level, say the N-level, is the one at which we are united with another or with more than one other. It is the one at which we lose the Self in emotional attachment to one other – and together we form a super-Self. The links between us are the binding cement of human relations; they vary from one such super-Self to another, but they effectively form a higher level of entities in the hierarchy. In the same way we can consider the next level, $N + 1$, as the one where we find collections of the N-level entities, as larger social groups such as clubs and societies, and where work-oriented sets of people are to be found. This identifies the workplace-Self, or the social-group-Self, which is at a higher level than that of the family or circle of loved-ones. At the next level we shall find even larger groupings; I have indicated the geographical region as a focus, but maybe just the town-Self would be as appropriate. Parallel to this will be the cultural activity of different sets of people – whose interests might be designated by single words like 'literature', 'theatre', 'music', or 'science', 'manufacture', etc. Then at the top level, $N + 3$, we would find the concept of nation-Self, covering the lower $(N + 2)$-level. It is clear that the sketchy table above is far from complete, but it is reasonable to suppose that it could be filled in with much greater detail, and possibly with one or more hierarchical levels into the bargain.

The possibility of a sudden drop from being an N-Self to being an $(N - 1)$-Self is ever present. The experience of *bereavement* is one such trauma, for in that moment, when the beloved is forcibly removed by death, the N-Self collapses to the loneliness of the $(N - 1)$-Self. When we come to discuss in more detail what I shall call the *geometry* of the inherent structure in a given relation (and by 'relation' I mean the cement which binds my beloved to me) we shall see that there is much more to be said about this sudden collapse of hierarchy, more detail associated with the notion of structure, but for the moment this experience (of bereavement) can be viewed as characteristic of the cause of tears. How much more joyful (full of laughter) is the experience of jumping upwards from being an $(N - 1)$-Self to being an N-Self? What is 'falling in love' if it is not that? What uplift to the heart, to the mind and spirit, is that event – and why does ordinary language insist on the word 'uplift' if it does not express, in some deep and commonly experienced sense, that hierarchical jump?

What about the anguish of unemployment – is that not the trauma of losing one's $(N + 1)$-Self? When the unemployed person feels that he/she has lost his/her 'self-respect', does that not mean that he/she – (what a nuisance all these he's/she's are, to be sure! Damn women's lib.) – has lost the respect which is the psychological awareness of being somehow 'superior' – that is to say, at a higher level? And if this level also contains various social-group-Selves, then we would expect the same cause of tears (sadness, despair, sense of defeat) when we are expelled from the group, from the club, when we are 'sent to Coventry', or otherwise rejected.

The concept of town pride, or of region pride, is an instance of the idea of $(N + 2)$-Self, whilst the concept of $(N + 3)$-Self manifests itself as 'patriotism' – particularly acute during times of war, for in such times the normal $(N + 4)$-level (which is super-nation-Self) is not available. All the citizens are then forced down into the $(N + 3)$-level and in order to stay there the idea of an $(N + 4)$-level must be actively fought against and rejected. But of course it may well be that some individuals cannot easily climb higher than the $(N + 3)$-Self. To them 'my country right or wrong', or common jingoism, is the highest ideal, the highest attainable condition. They cannot ever hope to understand those who insist that there is another level above that one. The latter people will be the natural opponents of the former and their enmity can easily lead to violence. Certainly it would seem that racism falls into the category of hierarchical limitations. I can only reject my coloured neighbour if I have no access to that super-cement which can bind us together and create a higher hierarchical level than the one in which I am aware of my white-Self (as the limit) and where I see him as the black-Self. That cement must eventually come from within me; my psyche must be able to move up that hierarchy by its own energies. Perhaps a little help can be useful from those who have reached it already, but that process needs careful handling; education is a dangerous process.

If we insist on pushing ever higher up the hierarchy we shall be searching after (and may never actually hope to achieve it) one ultimate level which unites all the lower ones. Every Self, at every level, will be then covered by the ultimate Self – if that can be attained. And is this not strongly suggestive of the Buddhist *nirvanha*, does it not correspond to the Christian ideal of searching for the condition of 'one with God'? In that condition, whatever we care to call it, we shall find all the peace of ultimate reconciliation because the absence of any hierarchical ceiling will mean that some super-cements will have been in such plentiful supply that 'rejects' will not have to be considered. The conflict of 'war' (at any level and between any 'Selves') will not be needed as it normally is if we are trying to tie ourselves down to some fixed-level Self. The prayers of the mystics have always expressed this condition – or the hope that striving may bring the suppliant to that condition. What better than a prayer by Francis of Assisi to this effect?

> Lord, make me an instrument of Thy peace;
> where there is hatred let me sow love;
> where there is injury, pardon;
> where there is doubt, faith;
> where there is despair, hope;
> where there is darkness, light;
> and where there is sadness, joy.
> Lord, that I may seek to console rather than to be consoled;
> to understand, rather than to be understood;
> to love rather than to be loved.
> For it is in giving that we receive;
> in self-forgetfulness that we find our true selves,
> in forgiving that we are forgiven,
> in dying that we are raised up to life everlasting.

I like to think that he had every intention of saying '... in N-Self-forgetfulness we find our true $(N + 1)$-Selves ...' The appeal to the Lord becomes an appeal for the psyche to reach the highest level in the hierarchy; achieving the Ultimate Self will automatically ensure the list of attitudes which go with all the lesser Selves. Less spiritually ambitious men than St Francis spend their time trying to achieve smaller steps on this psychic ladder. At the bottom end of our hierarchy we have placed the $(N - 1)$-Self and the N-Self and even here the move upwards from one to the other can be a difficult task. This move is certainly not linear, there being many aspects of the N-Self to be pursued. But perhaps the strongest one for the individual human is the unity involved in affection and love. Achieving this condition is an important part of psychic development and for most human beings might well be essential for the rest of the climb; for many it might be the zenith of their achievement. But rising above it (that is to say, rising above the N-level after achieving that through a personal love relationship) should not be confused with rejecting it. After all, the sense of uplift which comes with any hierarchical climb can induce such a heady feeling that we can be forgiven for being complacent at some modest achievement.

When Ben Jonson wrote his famous lyric poem, which begins

> Drink to me only with thine eyes,
> And I will pledge with mine;
> Or leave a kiss but in the cup
> And I'll not look for wine.
> The thirst that from the soul doth rise
> Doth ask a drink divine;
> But might I of Jove's nectar sup,
> I would not change for thine.

he was perceiving, in his experience of love, the sensation of hierarchical climbing which challenged the Ultimate Self. The 'thirst that from the soul doth rise' is the same as is found in the prayer of St Francis – which is why it asks 'a drink divine' – but for the poet his surrender in love is equal even to this.

But this achievement is so far not universal so that William Blake had to complain in one of his poems in Songs of Experience:

> Love seeketh not Itself to please,
> Nor for itself hath any care;
> But for another gives its ease,
> And builds a Heaven in Hell's despair.

(This is the Francis of Assisi theme again, characterizing the successful rise up the hierarchy via the experience of love, but then follows the 'experience' bit which is the negation of this rise:)

> Love seeketh only Self to please,
> To bind another to Its delight;
> Joys in another's loss of ease,
> And builds a Hell in Heaven's despite.

The selfishness of the $(N-1)$-Self makes it impossible to rise genuinely to the N-level; 'love' is then a hollow sham, being in fact a struggle between hierarchically bound $(N-1)$-Selves; we might even call it an $(N-1)$-war.

The loss of the N-level unity, for example by bereavement, is the other face of this coin, and the subsequent tears will be the antithesis of Jonson's exhilaration. Poets seem to find it very difficult to write about this experience – and not surprisingly so – but the following extract from a poem by Emily Dickinson seems to capture a great deal of the true sadness of this loss of structure:

I got so I could take his name –
Without – Tremendous gain –
That Stop-sensation – on my Soul –
And Thunder – in the Room –

I got so I could walk across
That Angle in the floor,
Where he turned so, and I turned – how –
And all our Sinew tore –

I got so I could stir the Box –
In which his letters grew
Without forcing, in my breath –
As Staples – driven through . . .

The 'Stop-sensation – on my Soul – and Thunder – in the Room' expresses the sudden hierarchical cut-off, the casting down into the (N — 1)-Self. Then a few lines later comes the immensely powerful 'And all our Sinew tore'. Notice it is 'our Sinew' and not 'my Sinew', the former being the experience as the N-Self. Then the sharp pain of the piercing Staples holding and confining the Self in the new constricted space.

3. The Dimensions of Things

We now come to the idea of the dimensions of our experiences, and to do this by examining *relations* between hierarchical sets of data. Somewhere in these relations (which we can denote by Greek letters such as λ, μ, etc.,) there is to be found whatever it is that we need to call the *structure* and, furthermore, as promised in Chapter 1, this structure is to possess an interpretation which is 'geometrical' in some acceptable sense, and this geometry requires the idea of multidimensionality for its description. So where is the connection between a relation, say λ, and this idea of dimension? How are we to grasp the concept of, say, a 20-dimensional structure, and is it *really* necessary to do so?

To answer the first question we need only appeal to the elementary training that most of us have had in geometrical matters – whether it has been in some formal sense, like when we study 'geometry' in school, or whether it has been just picked up by way of what is common sense in our dealings with the physical world. If we have had a formal education in geometry, however modest, then we have actually learned something about what is properly called *Euclidean* geometry – because the Books of Euclid, written somewhere about 300 B.C., have dominated the educational scene for over 2,000 years. But I do not intend to appeal to any knowledge of formal Euclidean theorems etc. in this discussion; the ideas we need are so basic that we are all likely to have absorbed them, rather like osmosis, during the usual lengthy acquaintance with simple sums concerned with measuring things. In Euclidean geometry these 'things' go by the names of point, length, area and volume, and they are *qualitatively* distinct. This sense of qualitative difference is a pre-mathematical experience in each of us – which is why we use the word 'qualitative'. The mathematician has provided us with a language which helps to *describe* these differences and this is all he has done; he has *not invented* the differences in the first place (although to people who scorn any form of thinking it might well seem that he has). In the works of Euclid these differences are related through the idea of 'distance' (mathematicians call Euclidean geometry a metrical geometry for this reason) and so one can be confused by learning that area is something obtained by multiplying two lengths together whilst volume involves three lengths. These peculiar ideas are not fundamental to the idea of dimensions and we do not need to appeal to them when we say, as we do, that a point is a 0-dimensional thing, a line is a 1-dimensional thing, an area is a 2-dimensional thing, and a volume is a 3-dimensional thing. All these statements are *descriptions* (only), in a specialized language, of the qualitative differences we have already experienced at the intuitive level. Notice that I am not saying that area and volume are different because they are of different dimensions. That would be putting the cart before the horse. The mathematical description uses words out of its own lexicon (where else should it get them from?) and in this case those words represent number, that is, zero, one, two, three. It is because of this that the description gets called *quantitative* and then the idea that something has become 'quantitative' becomes so corrupted that it is assumed to be non-qualitative. If some experience is described firstly in English and then in Chinese we do not normally think that it has thereby become changed in its essential nature or even in what it appears to be. Describing qualitative differences in a quantitative way does not make them anything less than they were before – provided the description does not contradict our experience of them. When 'dimension' is confused with metrical ideas (ideas dominated by the concept of distance) we can find the fundamental notion elusive, but quantitative descriptions need not be so dependent. In this book the idea of distance plays no important role and the 'geometrical' ideas we shall introduce will not be tied to any ideas of 'size' per se. For example the idea of 'rigid shape' is a metrical idea; it depends on the notion of distances remaining fixed. If we abandon those ideas then we give ourselves greater freedom in imagining shapes; they can be wobbly and sort of plastic. Freeing the mind from the ancient metrical geometry of Euclid has been a spiritual exercise for generations of mathematicians, but that fact has probably not yet seeped through to the rest of the world.

We can see that the qualitative differences which mathematicians describe in terms of dimensions can also be described in terms of a relation, say, δ (here we use the greek letter delta, δ, because the word dimension begins with a *d*). This relation is between two finite sets of things; let us call them X and Y and write them as follows.

$$X = \{\text{geometrical point-positions, viz., A, B, C, D}\}$$
and $$Y = \{\text{point, line, area, volume}\}$$

Our *experience* of these ideas of 'point' etc. only requires the use of four points in a suitable relation. This relation, δ, can be represented by the following matrix array of 0's and 1's, and the conventional picture of a tetrahedron, in Figure 4, reminds us of the essence of the qualitative differences.

δ	A	B	C	D
point	1	0	0	0
line	0	0	1	1
area	1	1	1	0
volume	1	1	1	1

This tells us that 'point', as a member of the set Y, is related to only one member of the set X, say, A. The idea of 'line' requires two members of X (like C, D), 'area' requires three and 'volume' requires all four. Of course we cannot take any old four points to express volume; they must not be coplanar, they must 'clearly' form a 'piece of volume'. But notice how tautological such a statement is. Since the qualitative experience of the set Y *must precede its description* we cannot in all honesty now try to *explain* this set of things in terms of that description. We have that apparent experience of the set Y built in us as reasoning beings, or do we?

Certainly we have an aptitude for learning but is it not, even so, a result of education, of having it pointed out to us when we are young and impressionable and then having it pressed home by repeated sums and illustrations? There is no harm in that of course, but there is harm in it if it means that we thereby lose the ability to understand that process – so that we then begin to believe that our 'taskmasters' the mathematicians have invented these diabolical subleties in both their form (the description of them) and their content (the intuitive awareness of them) and that everyone else must make a superhuman effort to grasp them.

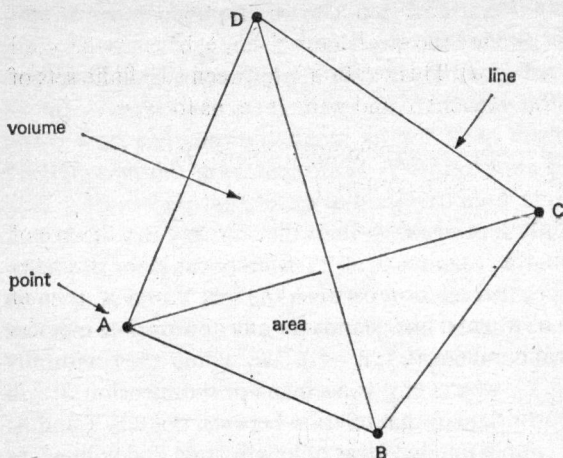

Fig. 4 Elementary dimensions

The introduction of dimensions is therefore being presented to you via a simple correspondence, as contained in the following description of the above set Y.

Name of member of Y	Number of members of X	dimension
point	1	0
line	2	1
area	3	2
volume	4	3

It is perhaps a bit irritating that the numbers in the dimensions list are merely one less than those in the middle column. But this is a historical accident and we must learn to live with it, otherwise we would need to unlearn all those reflexes which have been instilled into us.

Now if we have been badly educated we imagine that we are faced with the following problem: if I want to imagine something of 4, 5, 6, . . . dimensions what must I put in the set Y following the word volume? The answer to this is that the question should not be put, since it assumes that the whole process begins with that extreme right-hand column and proceeds to the left-hand one, whereas in reality the process has been the reverse of this. The question which can be put, and then answered, is the following one.

If I experience some entities in a set Y which are related to entities in a set X in such a way that (for example) a particular Y_1 (in Y) requires 10 members of X for its manifestation, then how many dimensions must I allot to this Y_1 *in the context of this information*? The answer is simple, even trivial: allot the dimension 9 to it!

We can be slightly more general than this and say that if, in that context, the member Y_1 requires $(q + 1)$ members of X for its identification, via the specified relation between the sets Y and X, then its dimension will be q (where q now stands for any appropriate member of the set of natural numbers, $0, 1, 2, \ldots$). We would then naturally speak of the thing Y_1, whatever it is, as being of q-dimension; it will have a q-value appropriate to that relation between the sets Y and X. In the case of our simple-minded ideas of length, area and volume, as they are related to our ideas of point-position, we would say that area has a q-value of 2 whilst volume has a q-value of 3. When the q-value (the dimension) of the thing we are interested in rises above the value of 3 mathematicians have tended to describe the thing as a 'hyper-volume'. But this is a lazy way out and not to be recommended, chiefly because it does not specify the actual q-value (and naturally this can vary a great deal) but also because it induces the feeling that the thing is still something like volume, but maybe more so. This latter idea is very destructive to that mental freedom which we shall require in pursuing this thesis. And this brings us up against that second question which we posed at the beginning of this chapter: how are we to grasp the concept of a 20-dimensional structure?

Now you will not be surprised if I claim that this must be replaced by a different question, which places the emphasis more squarely on the *descriptive role* which mathematics plays – as opposed to an imaginary causative role. This acceptable question will be: 'How are we to experience those relations $\lambda, \mu, \ldots$ between sets (like Y and X) which result in a description couched in the language of multidimensions?'

The short answer is that Life is full of them.

In the first place we have seen already that the hierarchical nature of data, via cover-sets, naturally provides us with relations. These in their turn have representations via matrix arrays and thence we find the dimensions (q-values) associated with the structure of that data.

But even so, such an example probably still seems to be somewhat abstract. So let us look closer at some of the ordinary things in life and see what extraordinary dimensions are to be found therein.

Looking out of my window into our garden (lovingly, almost obsessively, tended by my wife) I see it, such is my idiosyncratic tendency, as a collection of overlapping areas. These are called A_1, A_2, A_3, A_4, A_5, A_6 and shown in Figure 5 below. The boundaries of these areas seem visually natural, because of the curves and lines of the garden, and are identified by the following sequences of numbers on the diagram:

$$A_1 = (7, 9, 11, 7) \qquad A_2 = (10, 12, 14, 10)$$
$$A_3 = (1, 3, 13, 1) \qquad A_4 = (1, 4, 8, 9, 10, 14, 1)$$
$$A_5 = (2, 5, 6, 2) \qquad A_6 = (4, 5, 8, 4)$$

This set of areas is related to a set of garden plants which happens to be the following:

Set P: Red roses (denoted by RR), Yellow roses (YR)
Herbaceous plants (H), Shrubs various (SH)
Fruit bushes (F), Conifers various (C)
Deciduous trees (D), Lawn (L)
Annual bedding plants (AN), Vegetables various (V)

On examining the garden closely I notice that the relation, say λ, between the set of areas, A, and the set of plants, P, can be represented by the following array of 0's and 1's, where a 1 in the ith. row and jth. column means that Area A_i contains plant-type P_j.

λ	RR	YR	H	SH	F	C	D	L	AN	V
A_1	1	1	0	1	1	0	1	0	0	1
A_2	1	0	0	1	1	1	1	1	0	0
A_3	1	1	1	0	0	0	1	1	1	0
A_4	0	1	0	0	1	1	0	1	0	1
A_5	0	0	0	0	0	1	0	1	1	0
A_6	1	0	1	1	1	1	0	1	1	1

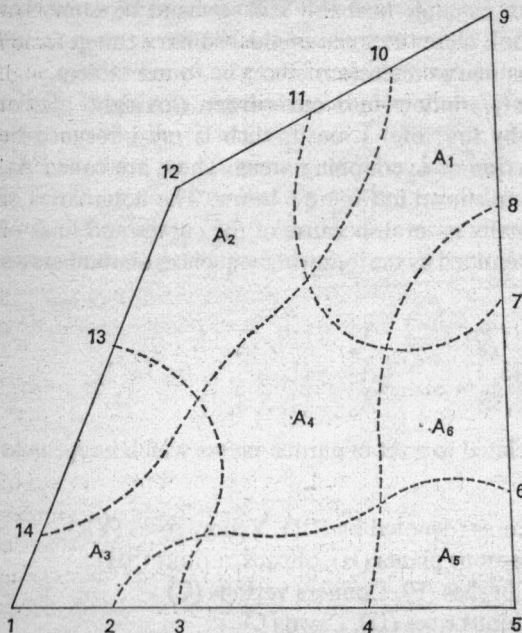

Fig. 5   A cover-set of areas for my garden

In terms of the set of *vertices* (the 'points' which are members of P) which make up the names of the columns of this matrix array, the areas of the garden have dimensions (q-values) as follows.

$$\dim(A_1) = 5 \qquad \dim(A_2) = 5 \qquad \dim(A_3) = 5$$
$$\dim(A_4) = 4 \qquad \dim(A_5) = 2 \qquad \dim(A_6) = 7$$

The area A_6 possesses the largest dimension ($q + 1 = 8$, so $q = 7$) whilst A_5 has the least. This expresses the fact that A_6 possesses the greatest *variety* of plants, whilst A_5 possesses the least; so A_5 is relatively monotonous (just as we might say that a Euclidean point is monotonous relative to a piece of Euclidean volume). The other areas are roughly equal; so in this case, with respect to the relation λ between garden areas and garden plant-types, dimension corresponds to one's sense of visual and horticultural variety.

But is this sort of dimension as real, as absolute, as the familiar ones of elementary geometry? Is it not more ephemeral, somehow less tangible and reliable than those old friends like area and volume? Well, the dimensions of these pieces of the garden are real enough since they arise quite logically from the relation λ given above, but of course they are not absolute, for various reasons. Firstly, because the relation λ is not absolute; it is perhaps peculiar to my own personal view of the garden, and is certainly not common knowledge since its properties have not been incorporated into our school curricula for the past 2,000 years. But, secondly, we can alter these dimensions by altering the set P of plants. This can be done most naturally by moving up the hierarchy in our view of the data. If we say that the set P is a set of N-level plants, let us now introduce a new cover-set, say P', which contains only one word, 'plants'. Then at this (N + 1)-level each piece of garden, A_i, is a 0-dimensional thing – because in the induced relation between the set A and the new set P' there is only one column (with a title of 'plants') and this is to be found in each of the areas. So from this higher level, N + 1, the dimensions of the A's have all been effectively reduced. Even so, the dimensions are still real enough in this context (of the relation λ). Nor should we imagine that the same sort of 'trick' cannot be (and is not) performed on pieces of our traditional geometry, when it suits our purpose. The Newtonian theory of planetary motion is expressed in a mathematical form in which even a thing like the Earth or the Sun is treated like a point (a 0-dimensional thing). If we go far away from an object, like a 3-dimensional house, we find it just as real to use a new relationship between things in which the house becomes a 'point'. A network of cities on a map is a structure which is at a higher hierarchical level than the one in which the cities are 3-dimensional, and on this network we confidently place numbers to represent distances between the cities. Ah, you say, but the objects are *really* 3-dimensional all the time – we are only pretending they are not for our convenience. To which the answer is that the same goes for my garden when I 'pretend' it is 0-dimensional at the (N + 1)-level. How do we ever know which is the basic rock-bottom level at which we *ought* to see the data and to examine the relations therein? There is no absolute hierarchical level, only more familiar ones. Given the data in the form of certain sets at specific hierarchical levels then, in that context, there is no 'pretence' about the dimensions nor is there anything unreal about them. It only seems that some relations are more ubiquitous, more familiar, more expected because of our educated and conditioned minds, than are others. But that might well be largely due to the fact that we stop looking after the education mill has exhausted our minds. The apparently basic 3-dimensionality of things, like the dimensionality of this book, is itself a consequence of relations between 'points' which are experienced via our sense perceptions of sight and touch. If we looked at this book by an X-ray device it would probably disappear altogether, and all the basic relations with it – not even 0-dimensionality would remain. Physicists believe in a particle called a neutrino whose properties are such that it travels right through the Earth without noticing it. What kind of dimensional world would we live in if all our perceptions depended on the agency of just such a particle? What would happen to our familiar notions of point, line, area and volume if the only experience of such things was via our sense of touch as transmitted through a feather duster? We are so busy educating the blind so that they can 'see' the world as we see it (conning them into believing that that is how it *is*) that we have missed a great opportunity to learn from them – as to how the world appears (geometry and all) without the perception of sight. A sympathy with that view of the world would help us to free ourselves from our own intellectual prison where our sensitivity to the reality of things is trapped in a banal geometry which imposes ludicrous limitations on the dimensionalities of them.

But to return to our simple garden–plants' relation λ and the N-level dimensions: why not turn it around and regard the A's as vertices and the plant-types as being related to them; in other words, look down the columns of the matrix array instead of along the rows? Now the relation gives us a new structure for the members of P and in this structure the dimensions of the members are 4 for Lawn, 3 for Red roses, Fruit bushes and Conifers, 2 for Yellow roses, Shrubs, Deciduous trees, Annual plants and Vegetables, and 1 for Herbaceous plants. This time the dimension represents our sense of the *spread* of any particular type throughout the garden.

This suggests that every relation, like our λ, between two finite sets (typically called X and Y) gives us two ways of looking at the data. Either we see the members of Y as defined by subsets of X, and thence they have appropriate dimensions (looking along the rows of the matrix array), or we see the members of X as defined by subsets of Y and then they have ascertainable dimensions (by looking down the columns of the array). If these particular sets are at the N-level in some hierarchy of data then the dimensions we find are appropriate to that level only. When we treat the Y's as the 'points' and look down the columns to find how the X's are defined we usually say that we are using the inverse relation – which is denoted by λ^{-1}.

As another example, suppose that Y denotes a set of named individuals, perhaps a hundred or more of them, and suppose that these people are related to another set X of medical symptoms – containing maybe fifty features, such as only doctors are capable of thinking of. The relation, λ, between these two sets is obtained by constructing a matrix array of 0's and 1's in the obvious way: if person Y_i shows the symptom X_j then we put a 1 in the ith. row and jth. column of the array. In the resulting structure the dimension of, say, Y_1 is a measure of the range of different medical symptoms which he exhibits whilst, looking at the inverse relation, λ^{-1}, the dimension of a particular symptom X_1 measures the set of people who suffer from it. A high dimension for X_1 means that the symptom is widespread among the people, whilst a high dimension for Y_1 means that he is severely ill (relative to a low-dimensional person), or that his illness is more interesting to the medical profession. The identification of specific illnesses in clinical diagnosis corresponds to moving up the hierarchy to a cover set, say X', of X – because, for example, influenza is a set of symptoms. Even so, we do not expect the cover-set to be a partition of X – for example, a common cold and influenza share the symptom of 'high temperature'. So the *sharing* itself *becomes an important feature of diagnostic classification*. This is why diagnosis can be wrong and why good clinical diagnosticians are doctors who have a feel for the overlap in the hierarchy of cover-sets, although they might not express it quite like that. It is also why we must search into this overlapping situation in order to find a way of handling it. A feeling that the possible overlaps in a cover-set make the whole thing just too complicated to deal with has often resulted in a search for partitions of the data. Taxonomies have been sought which express this idea of partition (that seductive gut feeling that if something is A it cannot be B), of forcing things into exclusive single classes, even at the expense of common sense. For example, the Central Statistics Office in dealing with census information about jobs uses the Standard Industrial Classification (at three levels) as a net to catch all employed persons. But the two upper levels it uses consist of successive partitions of the bottom level. So if you are a skilled lathe operator in light engineering you cannot also be a skilled lathe operator in shipbuilding, even though the small business you work for provides components for both of these categories. It is not just a question of whether or not we count people only once or twice or thrice etc. It is a question of whether we decide to separate people into different *dimensional* classes, into apples and oranges, and then count them separately. For if some lathe workers are of higher dimensions than others, in this sense, they must be of different significance in the structure of industrial society – as we shall try to explain in the sequel.

Describing the structural geometry

My good friend John Henry Doe lives with his wife Julie and their three children, William (aged 16), Sarah (aged 13), and Amanda (aged 7), in the suburbs of this English country town. They watch quite a bit of television in their spare time and tend to discuss the programmes in terms of a set X which has the following ten members in it.

Set X
X_1 = historical drama
X_2 = science fiction/horror films
X_3 = social comment documentary
X_4 = classical music/opera
X_5 = personality-cult pop show
X_6 = children's programme
X_7 = domestic situation comedy
X_8 = olde time music hall
X_9 = sports programme
X_{10} = news/current affairs

There is a relation between the members of the family and this set and it is determined by listing which programmes they like watching best. We shall denote the relation by the letters TV and, at the last time of asking, I was assured that the relation was described by the following array:

TV	X_1	X_2	X_3	X_4	X_5	X_6	X_7	X_8	X_9	X_{10}
John	0	1	1	0	1	0	1	0	0	1
Julie	1	0	1	1	0	1	1	1	0	0
William	0	1	0	0	1	0	0	1	1	1
Sarah	1	1	0	1	1	0	1	1	0	0
Amanda	1	0	0	0	1	1	0	1	0	0

This relation tells us something about the Doe family in the context of their television likes and dislikes. If that were the only information we could ever have about them it would define the observable limits of the world, or what mathematicians would rather call the *space*, in which they live. It tells us, for example, that William likes X_2 (science fiction/horror films), X_5 (pop artist shows), X_8 (olde time music hall), X_9 (sports programmes) and X_{10} (news/current affairs). These are indicated in the matrix array by 1's in William's row and in the appropriate columns. The relation also tells us, via the inverse relation TV^{-1}, that X_6 (children's programmes) are liked by (only) Julie and Amanda. This is obvious by looking down the X_6 column and finding where the 1's appear.

Now TV tells us that Julie is a 5-dimensional person in this context, because she is related to six of the elements of X. Now let us compare this with the situation illustrated in Figure 4 above, and by its accompanying table of old-fashioned geometrical things, 'point', 'line', etc. Because this is so familiar and because the geometrical words of Euclid have found a permanent place in our vocabularies, we look for a way of describing our new geometry in terms of the old. If we were lazy we might be tempted to just say that Julie is a 'hypervolume' – meaning something whose dimensions are greater than the familiar three of Euclidean geometry. But that will not do because it is not precise about the dimensions and also because it is too near the Euclidean case. Somehow it does not sound complimentary enough so to describe Julie; it makes her sound too lifeless, too much like a piece of wood (because our pre-mathematical experience of the Euclidean geometry is associated with lifeless inanimate things). We need a new word to describe that 'thing' which Julie has turned out to be and that new word must precede any of its descriptions (for example, the description by the matrix array); it should also have an associated dimension which is easy to remember. The word which mathematicians have used for this is *simplex*, together with its associated dimension. So we shall want to say that Julie is a 5-simplex, in this relation TV. This 5-simplex is determined by the six vertices $\{X_1, X_3, X_4, X_6, X_7, X_8\}$ but it would be wrong to say that Julie *is* this set of vertices. What she is, and what the word 5-simplex describes, is something that we can only know intuitively from the relation TV – just as, in the last resort, what we know about the word 'area' is intuitive in the context of relations between Euclidean 'points'. So, to keep this subtle distinction alive, we write the representation of Julie as

Julie = a 5-simplex in TV, denoted by the symbol

$$\langle X_1\ X_3\ X_4\ X_6\ X_7\ X_8 \rangle$$

and the use of the triangular brackets is the distinction between this and the *set* of these X's.

Returning to the Euclidean geometry of Figure 4 we would now say that in that relation δ, point = 0-simplex = $\langle A \rangle$; line = 1-simplex = $\langle C\,D \rangle$; area = 2-simplex = $\langle A\,B\,C \rangle$; volume = 3-simplex = $\langle A\,B\,C\,D \rangle$.

When we move from these forms of writing them down and think about the *geometry* of them (which is how we intuitively meet them in the first instance) we notice that each of them is represented by a convex polyhedron – with 1, 2, 3 or 4 vertices. These words can be taken over directly and we can say that, in the new geometry which represents a relation like TV, the X's act as 'vertices' of a convex polyhedron (which must lie in a suitably high-dimensional space) and that each simplex can then be described as being one such abstract polyhedron. Julie has a representation by a polyhedron with six vertices, John has a representation by a polyhedron with 5 vertices (he is a 4-simplex thing), and so on for the other members of the Doe family. But then we might well ask, 'What's so new about this geometry; can we not simply represent Julie by an ordinary octahedron (which has eight faces but only six vertices) in ordinary Euclidean space – which is still a 3-dimensional object in the traditional sense?' If this were so, then certainly the need for the polyhedron to be in some new higher-dimensional space which transcends the familiar Euclidean one would disappear.

The answer to this lies in the context of the relation TV and the

1. A low-dimensional view of woman – the sexual object

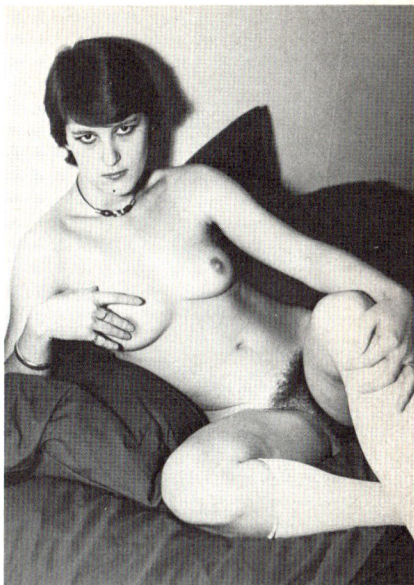

2. A naturalistic view of woman as a multidimensional structure

3. The Mona Lisa (Leonardo da Vinci)

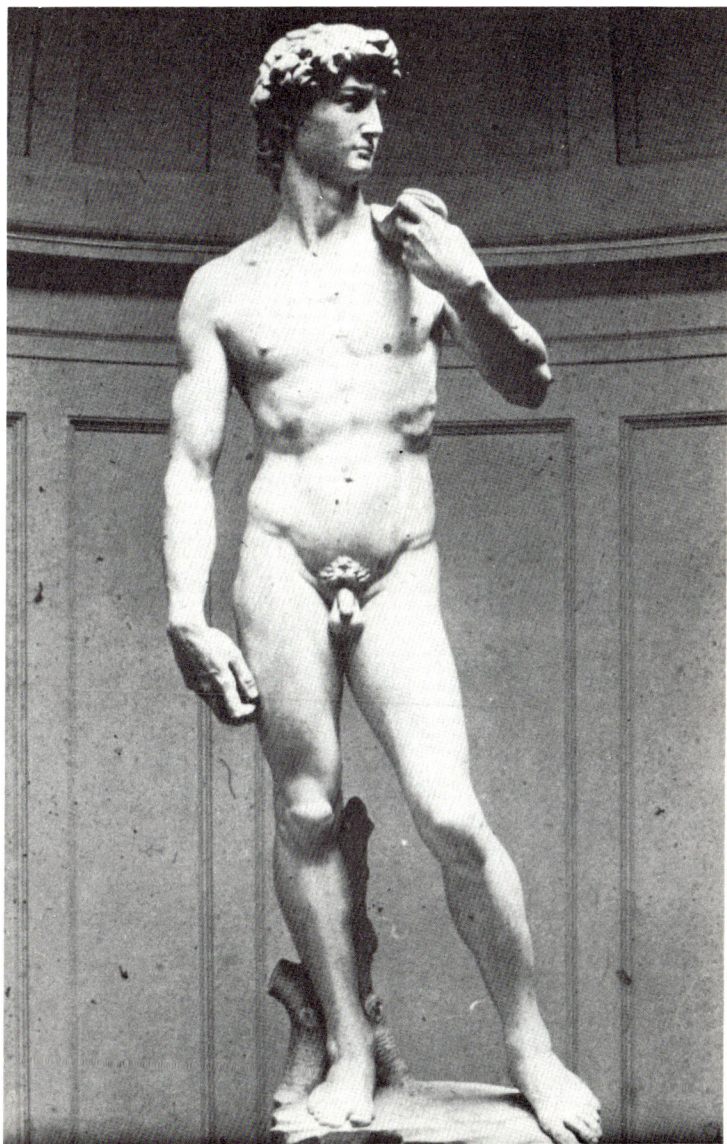

4. David (Michelangelo)

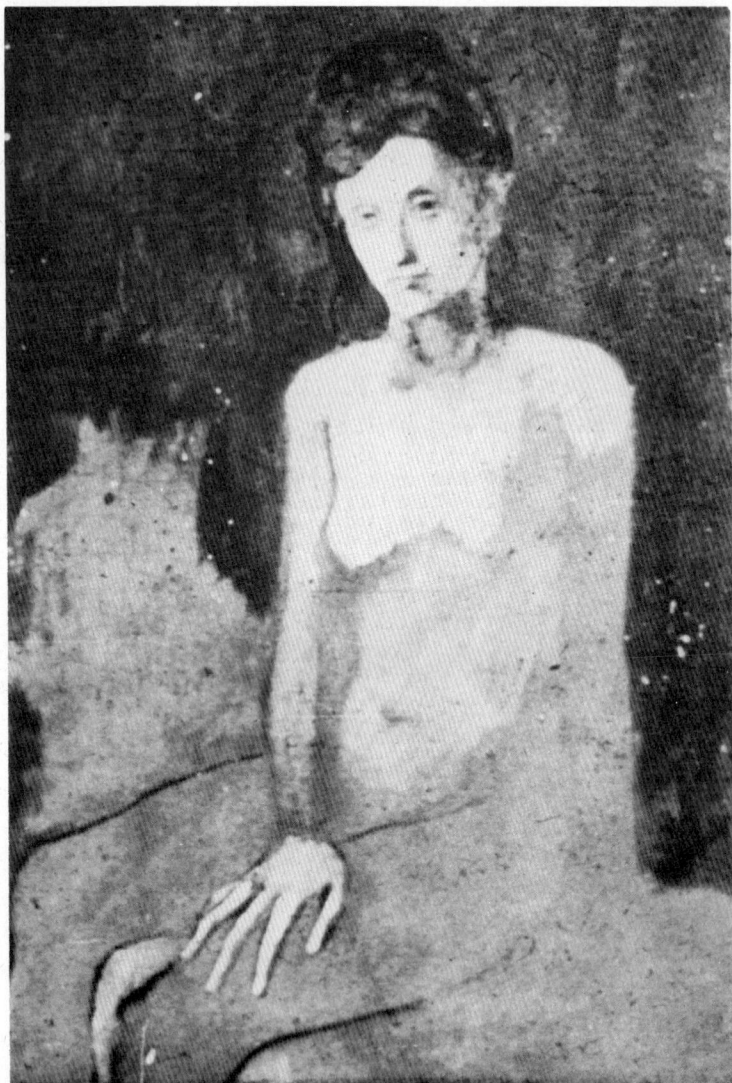

5. Seated nude – naturalistic (Picasso)

6. Seated nude – cubist (Picasso)

7. M. Vollard – naturalistic (Picasso)

8. M. Vollard – cubist (Picasso)

9. *(right)* Woman with guitar – cubist (Braque)

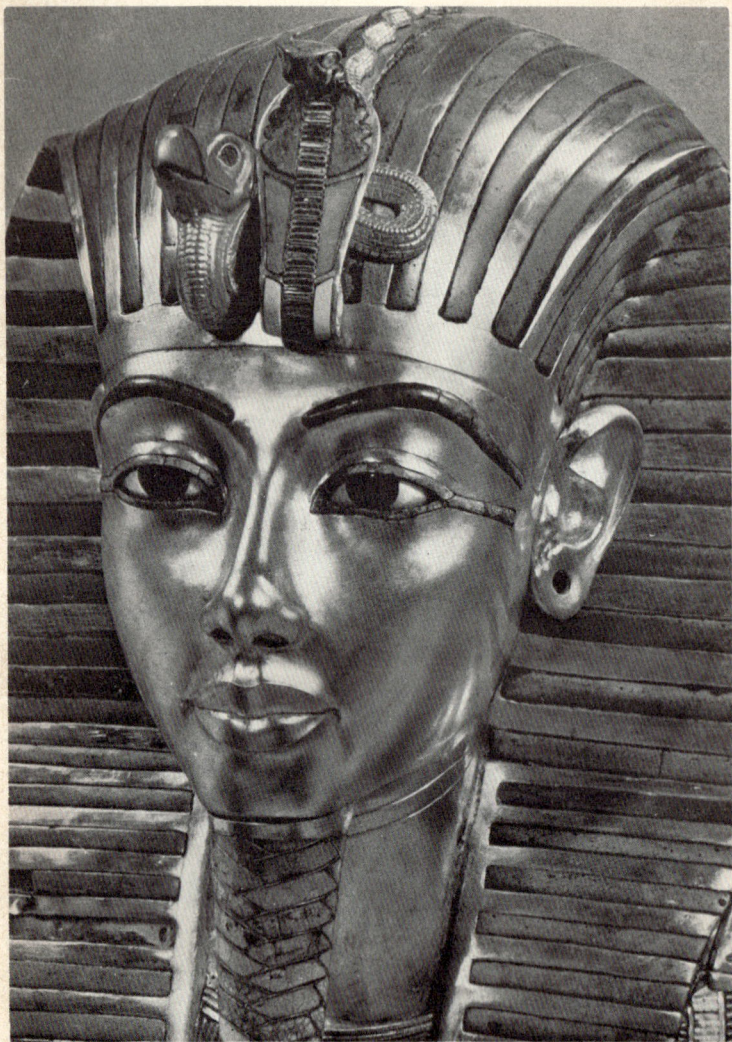

10. Mask of Tutankhamun – timelessness expressed by a p-event.

special key role of the vertices, the X's, in that relation. The point is that the data at our disposal *consist entirely and yet only* of the set Y, the set X, and the relation TV between them. We are regarding the set X as the 'vertex set' in terms of which the Y's are various p-simplexes (p taking values out of the numbers 0, 1, 2, 3, ...). In the resulting geometry the *only points* are these X's; there are no others available; the idea that a simplex is *represented* by a Euclidean-type polyhedron (in some vaguely-felt extension of our familiar Euclidean geometry) does not mean that that is what it is; all those 'extra' points lying about in the middle of a 2-simplex (a triangle with three vertices) are not really there because they are not mentioned in the data. For example, if we take the case of John Doe and notice that he is a 4-simplex in this relation TV, then we need a polyhedron with five vertices to represent him. Can we draw such a polyhedron in our familiar Euclidean space? Yes, of course we can, and it will look something like Figure 6. The trouble with that is that we notice that John, who is the 4-simplex

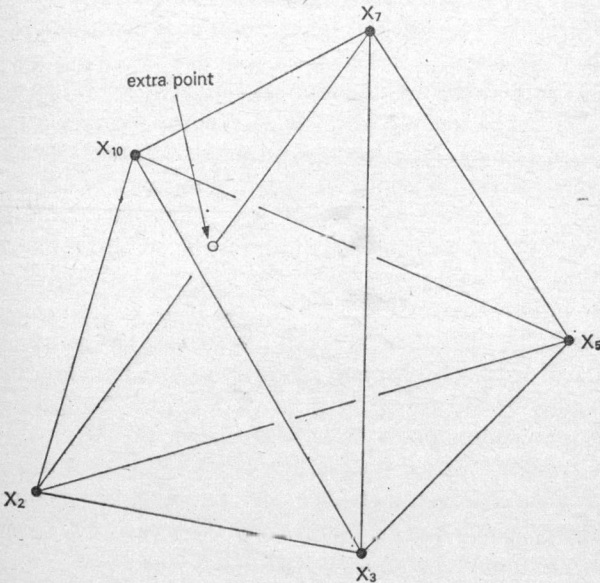

Fig. 6   Failure to embed John in ordinary space

$\langle X_2 X_3 X_5 X_7 X_{10} \rangle$, contains five tetrahedra within him, for example $\langle X_2 X_3, X_5 X_{10} \rangle$ and $\langle X_3 X_5 X_7 X_{10} \rangle$, and that these meet each other in the triangular face $\langle X_3 X_5 X_{10} \rangle$ – this is fine because this meeting can be adequately described by a subset of X, without having to invent new points in the geometry. This meeting of pieces of John is also an essential part of his (new) geometrical structure. Every geometrical property which John is to have, via this relation TV, must be seen to be part of the geometry of his representative 'polyhedron'; all the intuitive ideas of geometrical *connections*, of what is joined to what, of how things meet and intersect, must be so describable. All these properties must be found, and *none of them violated*, in the new geometry. But as far as John is concerned we notice that the simple-minded Euclidean figure of Figure 6 does violate the connections of his 4-simplex. The tetrahedra (3-simplexes) $\langle X_2 X_3 X_5 X_{10} \rangle$ and $\langle X_2 X_3 X_5 X_7 \rangle$ have an extra intersection because of the meeting of the edge $\langle X_2 X_7 \rangle$ with the triangle $\langle X_3 X_5 X_{10} \rangle$. This new 'point' is not one of the X's and so it cannot be part of the data; we must not recognize it as a point in this geometry. To accommodate this we need to spread out the figure somehow so that this intersection does not arise. This idea of needing 'extra space' is exactly the idea of needing an 'extra dimension'. So John needs a 4-dimensional space (we knew that already from TV) and so his own peculiar polyhedron-representation has to be one in a 4-dimensional space. For the sake of brevity we say that John's representation as a polyhedron is in the *space* E^4. Ordinary Euclidean space is denoted by E^3 and any p-simplex (involving $(p + 1)$ vertices) in any relation, λ, will need the space E^p for its accommodation. This is using an extended language of Euclidean geometry; the intuitive concepts must be found in the context of the relation under study – like the one about the medical symptoms already mentioned. So the peculiar new geometry that we need is determined in actuality by the matrix array which describes the relation. It is pointless to cudgel one's mind into imagining a 'genuine' extension of Euclidean space (unless that is appealing and brings some sort of intellectual satisfaction) – all the actual connections inherent in the geometry can be found from the array, and this is usually done by a computer programme, when it is done at all.

Now if we collect together the representations of John, Julie, William, Sarah and Amanda in the relation TV it is clear that their various simplexes are more or less connected up, because of the vertices they might share. The resulting geometry for the whole family is then a sort of connected set of polyhedra, each in its own special space E^p, and the totality of this is usually called a *complex* (or a simplicial complex, being a complex of simplexes). The space in which this complex must be sitting will not necessarily have to have dimension which is the sum of all the separate ones, because of the pieces they share. In fact it can be shown that if n is the largest dimension of the simplexes of the whole family then the totality of polyhedra which represents them can be accommodated in a space E^{2n+1}. For our purposes this is not very important, but it might be of interest to know that, using this approach of relations to study the game of chess, it is easy to prove that chess is a game described by a structured geometry in E^{53} (in 53-dimensional space) – no wonder it is a difficult game to play.[5] You can also demonstrate that E^3 (the case when $n = 1$, so that $2n + 1 = 3$) is necessary when playing the children's party game of joining up the three houses to the three utilities of gas, water and electricity – with no lines crossing. This cannot be done in the plane (that is to say, in E^2) but requires Euclidean three dimensions, E^3, for its solution (what you are trying to draw is a collection of 1-simplexes (the lines which join a house to a utility) which constitute a complex where $n = 1$).

If we return to our relation TV with the Doe family, we notice that Amanda is only a 3-simplex, i.e. Amanda $= \langle X_1 X_5 X_6 X_8 \rangle$, and this can be easily represented in E^3, being a tetrahedron. All the features of Amanda can now be 'seen' in the ordinary sense and these features consist of all the sub-polyhedra (smaller simplexes) which go to make up the total one. In this sense Amanda consists of the following parts:

(i) the four vertices (0-simplexes) $\langle X_1 \rangle$, $\langle X_5 \rangle$, $\langle X_6 \rangle$, $\langle X_8 \rangle$

(ii) the six edges (1-simplexes) $\langle X_1 X_5 \rangle$, $\langle X_5 X_6 \rangle$, $\langle X_6 X_8 \rangle$, $\langle X_1 X_6 \rangle$, $\langle X_1 X_8 \rangle$ and $\langle X_5 X_8 \rangle$

(iii) the four triangular faces (2-simplexes) $\langle X_1 X_5 X_6 \rangle$, $\langle X_1 X_5 X_8 \rangle$, $\langle X_5 X_6 X_8 \rangle$ and $\langle X_1 X_6 X_8 \rangle$, and

(iv) the one tetrahedron $\langle X_1 X_5 X_6 X_8 \rangle$.

Each such contribution is a simplex in its own right and is represented by an easily visualized convex polyhedron in E^3 – being various obvious pieces of an ordinary tetrahedron.

It is natural to call any component simplex of a larger one a *face* of the latter. Thus Amanda exhibits 15 faces, if we allow that the tetrahedron is a face of itself (just to be tidy). So which face does she turn to her mother Julie, in this context? Why, the face they share, of course, and that is the simplex (or 'triangle') $\langle X_1 X_6 X_8 \rangle$. Using the natural language of geometry we would then want to say that, with respect to this relation TV, Julie and Amanda are *connected* via this face – and only by this face. Because this connection is through a 2-dimensional face (a 2-simplex) we would naturally call it a 2-*connectivity*. If they happened to share a q-simplex we would say they are q-connected. If this relation is the only thing which connects them (which of course it is not) then the geometry of that relation tells us that they can only communicate through this common shared face of their respective abstract polyhedra. The only things they can talk about will be combinations of the vertices X_1 (= historical drama), X_6 (= children's programmes) and X_8 (= olde time music hall). In the total geometry of the relation TV, in that simplicial complex formed by the whole family, the Doe complex, this special 2-dimensional space E^2 is where they meet – and they do not meet anywhere else. So the general idea of connectivity of that complex, numerically specified as it can be, tells us how the family is joined up to form a whole – and also how it is not connected. The possibility of meaningful communication between the members of the Doe family is therefore constrained by the connectivities of the geometry which represents their relations, and this geometry is multidimensional in some E^n (in this case the value of n is not greater than 5, because the largest dimension is 5) (see Sarah's line in the matrix on p. 79). This new geometry would be the *backcloth* against which their mutual actions would have to take place; there would be no other. And although this is not the whole story (because there are many more relations which connect the members together) yet it is typical (paradigmatic even) of *how that story can be told*. Through their common face (geometrical connection) Julie and Amanda can communicate, can know each other, can exchange concepts and attitudes. Of course this communication must be about things which depend entirely upon these vertices $\{X_1, X_6, X_8\}$; no other kind of *traffic* can exist on that structured backcloth. Because the channel between them is a 2-dimensional thing we shall want to say that the most it can carry is 2-dimensional traffic (abbreviated to 2-traffic). This means simply anything which, at the most, simultaneously involves the three concepts which are represented by the vertices X_1, X_6 and X_8. So they find their rapport, in this context, via their common attachment to historical drama, children's programmes, and olde time music hall. They do *not* discuss classical opera or sports because the geometry which faithfully represents them does not permit this – there is no connectivity between them in those areas.

If we let Y represent the set of people called the Doe family, then the simplicial complex we are talking about can be denoted by the symbol KY(X), where K reminds us of the hard *c* in complex, Y is the set of simplexes we are interested in, and X is the set of vertices in terms of which those Y's are determined via some well-understood relation like TV. This notation is particularly neat because we can naturally turn it around and write KX(Y) when we are referring to the other associated complex, the one which is derived from the inverse relation TV^{-1}. In this latter case the Y's act as the vertices of the geometry and the X's are specified as certain simplexes more or less connected together by way of the inverse relation. Since these two complexes are so closely associated (although it would be foolhardy to assume that their peculiar geometrical properties are in any way the same) we commonly refer to them as being *conjugate* to each other: we say that KY(X) is the conjugate of KX(Y), and vice versa.

In the structure KY(X) Sarah is a 5-simplex and she shares a 2-dimensional face (which happens to be $\langle X_2\ X_5\ X_8 \rangle$) with her 4-dimensional brother, William. On the other hand, in the conjugate structure KX(Y) we can see that X_2 (= science fiction/horror films) is a 2-simplex (being in fact $\langle$John, William, Sarah$\rangle$) which is itself a 2-dimensional face of X_5 (= personality-cult pop show) because the latter is the 3-simplex $\langle$John, William, Sarah, Amanda$\rangle$. (These facts all follow from the matrix array which represents TV.) So what kind of traffic would we expect to find on this conjugate complex; what can flow between X_2 and X_5 in this geometry denoted by KX(Y)?

Whatever that traffic is, it must be something which is determined by the vertices of their common interface – which is the 2-simplex $\langle$John, William, Sarah$\rangle$. An obvious possibility is the traffic of 'giving priorities to the programme watched by the Doe family'. The question of whether an X_2-programme or an X_5-programme should have some priority (like being first or second choice) is a matter to be decided by the Doe family members (who of course are the vertices in $KX(Y)$). These two types of programme have a common face which represents a family 'pressure group'. It consists of the subset {John, William, Sarah}. So the traffic which can flow, between X_2 and X_5, across this connecting face (this pressure group) is the traffic of 'priorities being altered between these two types of programme'. The people John, William and Sarah are the only people who can do the sort of internal/mental horse-trading between X_2-priority and X_5-priority − because of this place they occupy in the geometrical connectivity between X_2 and X_5. The priorities given to X_2 cannot even be discussed by Julie or Amanda because neither of them is in the X_2 simplex. Amanda cannot compare an X_5-priority with an X_2-priority because the latter is not related to her, although the former is.

Some idea of how we can view the complexities of family life can be gleaned from this discussion. If there are many relations between the members of the Doe family, which we are denoting by the set Y, and various other sets, which might be denoted by X, A, B, etc., then the whole set of structures like $KY(X)$, $KY(A)$, etc., and their conjugates $KX(Y)$, $KA(Y)$, etc., will provide a rich backcloth for relevant traffic between them. These various structures will naturally represent the varied multidimensional spaces (at various hierarchical levels) which provide the connections, the links, the social cement, between the individuals. Furthermore, each person will be a composite of many simplexes, will be a simplicial complex in his/her own right. (Look ahead to Figure 7 for a modest illustration of a complex.) For example, Julie's set of vertices taken out of only a couple of dozen relations could run into hundreds; her dimensions and connectivities with other individuals and/or other groups of individuals will thereby be expressible in the new geometrical structure which faithfully represents her as a complex of simplexes − denoted perhaps by K(Julie). Some people will enjoy greater rapport with Julie than will others. Some people will be totally disconnected from Julie, whereas they might well be highly connected (through a high-dimensional interface) with John, or with William or Sarah. Why else should Julie want Sarah to 'bring her friends home', if not to keep up the connectivities within the family, to make sure that the Doe geometry can carry all the potential traffic which its members might encounter outside of it? Making connections outside the home means just that – extending the geometry of one's complex, K(ME), and of course such connections introduce new channels for traffic flow, opening new horizons (literally) and letting in the traffic of new ideas and strange new attitudes – strange because they might well be determined by new sets of vertices which the home is unaware of. Surely it is inevitable that the family backcloth cannot be permanent, particularly for the younger members, but must constantly be under pressure to change. To the young the apparently 'rigid attitudes' of the old can only be understood in terms of a static backcloth – even a shrinking one. The ability to 'take in new ideas' must surely depend on the ability to accept and to grasp hold of new vertices in the K(ME). Furthermore the subtleties of the connections in a multidimensional space, like E^n, make it very easy for us 'to miss each other' in our attempts at making contact or, if not that, certainly to make less than adequate contact – perhaps only through a 2-dimensional face when we really need a 20-dimensional one to carry the 20-traffic which is in us. One way of making inadequate contact, and kidding ourselves that it is adequate, is by the device of mentally moving up the hierarchy. For this enables us to cut down on the number of the vertices and thus make it appear to be of smaller dimension; it is easier to make contact in an E^3 than it is in an E^{30}. Perhaps we begin to discuss the rights and wrongs of some specific action like 'should we give permission for Dr Drug, at the local hospital, to turn the switch and disconnect the life-preserving machine to which Aunt Mary is presently joined?'. The geometry we find ourselves in is the family one where K(Aunt Mary) is connected to you and to me and to all the others we know. This geometry is of very high dimension and when we discuss the problem with Dr Drug (who lives in a different geometrical space, K(Drug)) we find it difficult to 'get through to him' on the essential issues (which are geometry-specific) – just as he gets exasperated at failing to get through to us. The traffic is flying about through some dizzy spaces and chaos reigns in the discussions. Then perhaps we all discover that it is 'a matter of principle' – in other words, we can all move up the hierarchy to consider a new geometry where the dimensions are much smaller. All the Aunt Marys of this world have suddenly become a single (collective) person known as 'patient in terminal care' and we throw away our connections at the lower-level geometries because we find that at this new level the space is so small (dimensionally) that we can communicate; the question suddenly has meaning and is quickly settled; moving up the hierarchy has made cowards of us all. How else to explain how easy it is to pass by the beggar with the thought that 'the State makes provision for such deserving cases through its social security services, so I need not do anything about it personally – after all, I pay my taxes'? So as an $(N + 3)$-Self, connected to the State by a 0-dimensional channel which carries my taxes, I *fail to see* the $(N − 1)$-Beggar. If I dropped down to my $(N − 1)$-Self I would have to see the rags, poverty, illness and despair of that individual and then to decide if the basic K(ME) could carry some charity across our interface to the K(Beggar). That decision would probably have to be high-dimensional traffic and might require some $(N − 1)$-Self-knowledge which would be painful to admit. So I can now enjoy the luxury of being generous with the cowardice of being mean – just by moving up the hierarchy – the stuff that politics is made of.

Sorting out the connections in the geometry

In Figure 7 we have the geometrical representation of a typical complex. Because we cannot draw anything of dimension higher than three this figure cannot represent a complex with any higher dimensional simplex in it than that but, allowing for this, it is quite general and exhibits different levels of connectivity.

Let us suppose that in fact this structure is relevant to the working environment in which John Doe finds himself. He works as a business executive in a firm which deals with electronic and computing equipment, largely as an entrepreneurial enterprise, and shares various responsibilities with his five senior colleagues. The relation at, say, the $(N + 1)$-level of data which provides the structure is the following one, Θ, where the vertices in the X set can be described in the following way:

Set X $X_1 =$ personnel matters and working agreements
$X_2 =$ company accounts, staff salaries and wages
$X_3 =$ outside firms of suppliers, hardware and software

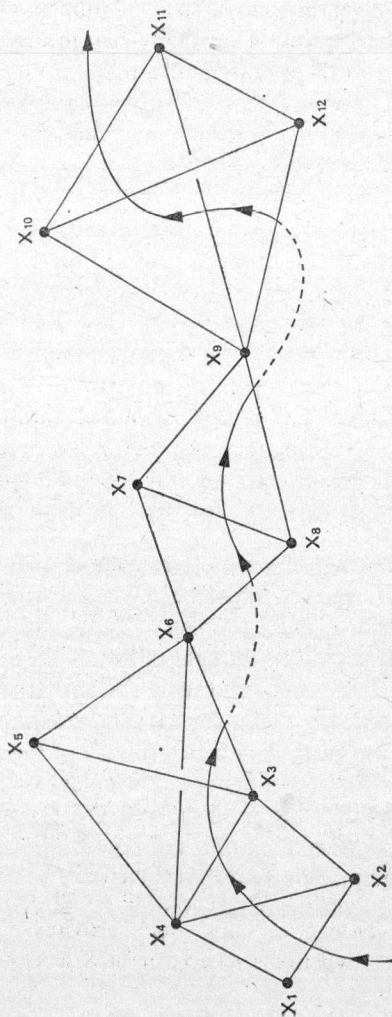

Fig. 7   The geometrical connectivities of a typical complex

X_4 = sales promotion and advertising
X_5 = transport and shipping matters
X_6 = firms which manufacture specific finished products
X_7 = sales outlets at home
X_8 = sales outlets abroad
X_9 = current account banking and Inland Revenue
X_{10} = finance houses and capital investment
X_{11} = insurance matters
X_{12} = shareholders and stockbrokers.

The set Y is the set of executives, of which John Doe is represented by Y_3 whilst Y_6 is the managing director, Richard Roe.

Θ	X_1	X_2	X_3	X_4	X_5	X_6	X_7	X_8	X_9	X_{10}	X_{11}	X_{12}
Y_1	1	1	0	1	0	0	0	0	0	0	0	0
Y_2	0	1	1	1	0	0	0	0	0	0	0	0
Y_3	0	0	1	1	1	1	0	0	0	0	0	0
Y_4	0	0	0	0	0	1	1	1	0	0	0	0
Y_5	0	0	0	0	0	0	1	1	1	0	0	0
Y_6	0	0	0	0	0	0	0	0	1	1	1	1

In looking at the figure we must remember that, from the above matrix array of 0's and 1's, John Doe (Y_3) is the 3-simplex $\langle X_3 X_4 X_5 X_6 \rangle$ which is a tetrahedron in the geometry – and not just the edges of it; so that tetrahedron should be thought of as 'solid', as should the other one $\langle X_9 X_{10} X_{11} X_{12} \rangle$ which represents Richard Roe (Y_6); and all the other triangles are also 'solid' for the same reason.

In this structure every member of this set of business executives is either 2-dimensional (occupying a 2-simplex, or solid triangle) or 3-dimensional (occupying a 3-simplex). The dimensionality of each executive expresses the range of his responsibilities – the things which he must feel simultaneously responsible for. The geometry shown in the figure is representative of the complex KY(X), of course, and so the way that pieces of it are connected (through shared sets of X's) is the way that the executives are connected (through their shared responsibilities).

At the level of 3 dimensions (at the level of q = 3, let us say) there are just two pieces in the geometry. They are the two tetrahedra representing John Doe and Richard Roe, and they are not directly joined at all.

So any 3-dimensional traffic (like business matters which are concerned simultaneously with four of the X's) must either exist in the John Doe piece of the geometry or in the Richard Roe piece. The common example of such traffic will be just those matters which require decisions of a ranking nature (like the allocation of resources in men or money) between four simultaneous areas of responsibility. This kind of traffic (its dimensionality) is exactly the kind that distinguishes Doe and Roe from the rest of the executives. But it means in general that the *number* of 3-dimensional pieces (we shall call them *components*) of the geometry of $KY(X)$ seriously affects the *mobility* of 3-traffic in the structure. The mobility of 0-traffic (matters which are only concerned with one vertex, no matter which) is not affected in this geometry because it is possible to move from one piece of it to another (however distant) provided we only need a joining through a single vertex. So we say that there is only one component at the level of $q = 0$; this corresponds to our first intuitive idea that the geometry in the figure is 'obviously connected' and in one piece. At intermediate values of q the number of components in the geometry varies. For example, at $q = 2$ (the level at which we see the 2-dimensional triangles – or, rather, the level at which 2-dimensional traffic is trying to exist in the structure) there are six separate components in the geometry. This is because each of the tetrahedra provides us with one component (all the triangular faces of a tetrahedron are actually connected to each other at an adequate dimensional level – i.e. through the middle of the tetrahedron, which is a 3-dimensional level) and each of the other triangles is a separate component (because none is joined to another by a 2-dimensional simplex (a triangle) – but only by an edge (like $\langle X_2X_4 \rangle$) or by a point (like $\langle X_6 \rangle$ or $\langle X_9 \rangle$). But at the 1-dimensional level ($q = 1$) where pieces need to share only edges to be connected, we have three components in the geometry. These are the pieces which represent the executives Y_1, Y_2, Y_3 – which form one component at $q = 1$, or Y_4, Y_5 – which represent another component at $q = 1$, or Y_6 – which represents the third component at $q = 1$.

If we fix our minds on any particular q-value, then we are saying that the simplex (the Y's in $KY(X)$) which form one particular component at that level do so because it is possible to find a path from one such simplex to any other (in the same component) which can be traversed by q-dimensional traffic. The numbers of these components, at the various q-values, are a first rough guide to how much *obstruction* to q-traffic is inherent in the geometrical structure because, as we have seen, if there are two pieces at some specific q-value then q-traffic cannot get from one to the other — *there is not enough geometry* (of the right kind) *to carry that traffic about*. A list of these numbers for figure 7 is the following:

dimension q	=	3	2	1	0
number of components	=	2	6	3	1

In this case we notice that John Doe and Richard Roe lead relatively lonely existences *at the level of* q = 3, because they are not connected at that level to anyone else: so *it is lonely at the top*. But everyone else is lonely at the level of q = 2, because each is a separate component at that level: so it is lonely being an executive anyway. But they come together, to some extent, at the level of q = 1 since they then form three distinct components. At this level John Doe has a small entourage (or confreres) who share connections which make them into one component. On the other hand Richard Roe (Y_6), the managing director, is still disconnected: so it is even lonelier being managing director.

The extent to which an executive is *relatively* lonely, because of the specific structure which relates them all, can be given a simple numerical value which seems to work quite well. We call this number the *eccentricity* of the executive (or of the simplex which he represents in KY(X)) and we define it in the following evocative way. First we identify his specific q-value (John Doe's value is 3) which we then call his *top-q* value; then we identify his *bottom-q* value, which is the largest q-value at which he is connected to any other executive. His eccentricity in the structure is then simply the number (top-q — bottom-q) ÷ (bottom-q + 1). (This precise way of finding a thing called 'eccentricity' is not sacrosanct, but it seems to evoke highly relevant intuitive feelings.)

In this imaginary case we find the following eccentricities for each of the executives:

Executive	Eccentricity	Executive	Eccentricity
Y_1	$1/2 = (2 - 1)/(1 + 1)$	Y_4	$1/2$
Y_2	$1/2$	Y_5	$1/2$
Y_3 (John Doe)	$1 = (3 - 1)/(1 + 1)$	Y_6	$3 = (3 - 0)/(0 + 1)$

So Richard Roe is the most eccentric (he stands out most among his colleagues in this structure), whilst John Doe is the next most eccentric. The curved line in Figure 7 shows the fact that 1-dimensional traffic on the structure can only exist in the three separate components, at $q = 1$. The dotted portions indicate where the traffic cannot flow, because of the lack of suitable connectivity in that region of the geometry.

We notice from the way that eccentricity is defined that when $q = -1$ the value becomes infinite, and this value of q (the weird 'dimension' of -1) is conventionally the value we give when there is no connection at all. If the diagram in Figure 7 fell into two quite distinct pieces, so that it could be drawn on two separate bits of paper, then we all agree (yes, we do) that they are said to be connected at the level of $q = -1$. So to be *infinitely eccentric* is to be totally disconnected from your peers – a sort of contradiction in terms – but it makes sense, because how much more eccentric could you then be?

We also notice that when your top-q equals your bottom-q, then the value of your eccentricity is zero. This means that you are 'the face of' someone else and you do not have any individuality in the structure – eccentricity is a measure of the individuality of the simplex. Each of us is eccentric, I trust, and each of us is burdened with the conflict between trying to increase our top-q whilst at the same time trying to keep an acceptably high bottom-q (in order to remain a member of the group). The 'drop-out' is presumably someone who has decided to make his bottom-q equal to -1 (?), but I doubt if he succeeds since he immediately becomes connected to other drop-outs, and his bottom-q begins to rise again. A large bottom-q means that we are highly connected to our fellows and that, together, we can form a geometry which can carry relatively high-dimensional traffic. A large top-q means that we have large dimension in the structure and that we can carry the relevant high-dimensional traffic (alone). To have a high eccentricity means that one is relatively unapproachable, that communication must be in terms of relatively low-dimensional traffic (using ideas which involve only a small number of vertices in the set X), and that one can also carry relatively high-dimensional traffic which *cannot be shared* with one's fellows. So the idea of eccentricity, quite apart from its arithmetical detail, corresponds to the idea of individuality conjoined with the idea of being one of the herd. Never to have an idea which is your very own, or to be active except at someone else's instigation, is to exhibit zero eccentricity in the structure. Standard military training is an example of trying to achieve zero eccentricity in the soldier; clearly the uniform is a simple outward sign of zero eccentricity. If Gertrude Stein had said (maybe she did) 'a soldier is a soldier is a soldier', it would have meant that you cannot find any non-zero eccentricity about a soldier, however many times you look at him or describe him. To exhibit zero eccentricity is to be only a face of someone else – you might as well not be there as far as the structure is concerned, you do not add to the geometry in any way – the structure does not need you because you are only the face of some other part of it. Surely this is an experience of adolescence (and beyond ?) – growing up brings the feeling that one has an individuality which is not just a face of that which is constantly being offered by one's parents or teachers. The unthinking (or is it unfeeling ?) parent sees his child as a face of himself – and that 'himself' is what he currently is, not even what he once hoped to be. The parent, or the teacher, might well be of zero eccentricity – what then for the individuality of the child or pupil ? As Sarah Doe grows physically and psychologically she finds her top-q value increasing and her eccentricity rising about the level of zero. Then probably she cannot share the restrictions of that traffic which her parents determine – she becomes eccentric, an individual, a 'person'. Perhaps little Amanda is still enjoying the delights of exploring the 'faces' offered her by her mother and father and teachers and wallowing in zero eccentricity. Perhaps 'innocence' in the young child is the expression of exploring through the low-dimensional pieces of the adult structure ? Surely it means more than that, but even so that is probably the backcloth against which it is set. The rise of eccentricity then becomes associated with what Jung calls the process of *individuation*. Nor is this necessarily in conflict with the rise of bottom-q, which corresponds to the process of social *integration*, because eccentricity is a relative measure. But society must somehow find a balance between these two tendencies, must decide how much eccentricity is acceptable relative to a given bottom-q. Such questions can involve actual membership of a society or a group, or a nation.

An example of structure from clinical psychology

This idea of a personal structure which expresses the concept of the individual is currently being applied in clinical psychology. One of the psychologists most active in this work is David Mulhall (who is currently (1979) working at the Middlesex Hospital, London) and the following anonymized case history is taken from his experience. In orthodox circles the novelty of this approach lies in the belief that individuals must not be treated as if they are merely statistical entities with various probabilities attached to them. So one would not wish to place any emphasis on a statement like 'his chance of having cancer is 1 in 5'. For if we have some hard scientific knowledge about this particular person and about how his medical geometry is related to that of the thing we call cancer then, for him, there is no question of 'chance' – either he has it or he does not. The question of whether or not he will suffer the illness in the future will then be a question of how his geometry can change to fit the case. Because this attitude places severe demands upon us it is very easy to become (scientifically) frightened and to take refuge in the non-knowledge of 'the chance of . . .'. In Mulhall's work this concern with the individual is expressed in terms of this structural geometry so that he has been able to open up new vistas in clinical psychology.

The patient was called John, let us say, and was in his mid thirties. He had been referred by his G.P. for treatment, suffering as an exposeur and with accompanying marital and other symptoms. His history included an emotionally cold childhood and more than one period in his life when he had been enuretic. But fortunately he was an intelligent man, which enabled him to co-operate with his treatment and to see a great deal which was explained to him in broadly structural terms (though not in the precise terms employed in the analysis discussed here). The result of this was a happy outcome for John and for his wife and domestic scene generally.

Discussions with the patient led to the identification (jointly with him) of seven mental states for himself and another set of seven mental states for his wife. The question of different hierarchical levels was not pursued by the clinician, so we can merely think of these vertices as being at some fixed N-level, for our purposes. The sets are as follows:

Husband's set, Y	*Wife's set*, X
Y1 = dominating	X1 = frightened
Y2 = aggressive	X2 = supportive
Y3 = affectionate	X3 = bitchy
Y4 = frustrated	X4 = inhibited
Y5 = protective	X5 = happy
Y6 = gentle	X6 = impulsive
Y7 = frightened	X7 = kind

During the treatment the patient produced a series of relations between these two sets, in the following way. He noted that, for example, when his mood was that of being dominating then his wife's states could be ranked (by numbers 0 to 6) in the order in which they seemed to him (from experience) to be a reaction. The zero meant that such a state (one of the X's) never transpired, whilst a six meant that it was most likely to do so. The result of this would be a matrix array in which the Y's would denote the rows and the X's the columns, and in which each row would have the numbers 0,1,2,3,4,5,6 in *some* order. By treating the numbers 4,5,6 as a 1 and all the others as a 0, a new matrix array was produced by the clinician, and this represented the most likely relation between John's moods and his wife's reactions. The structure of this relation we denote by $KY(X)$ and $KX(Y)$, as usual, and it is the structure between *husband's actions* and *wife's reactions*.

In the same way the patient would consider the inverse situation and produce a ranking of his own moods as likely outcomes when his wife's moods were regarded as the 'actions'. The same process of 'slicing' the rankings (between the numbers 3 and 4) gave a relation whose structure, $\tilde{K}X(Y)$ and $\tilde{K}Y(X)$, now represented the relation between *wife's actions and husband's reactions*.

As these structures were obtained and translated back for John to think about, and possibly to correct on his own initiative, the changes mirrored the progress of the treatment. Those changes are indicated by three separate occasions listed below.

Week 1:

In $KY(X)$ – husband's actions/

dimension, q	components (pieces) of geometry
2	(Y1,Y2), (Y3), (Y4), (Y5), (Y6)
1	(Y1,Y2,Y3,Y4), (Y5,Y6)
0	(all Y's)

In KX(Y) – wife's reactions/

dimension, q	components of geometry
3	(X1) (X4), (X7)
2	(X1,X3,X4), (X7)
1	(X1,X3,X4,X5,X7), (X2)
0	(all X's)

These tell us that Y1 and Y2 are the same (identical) simplex at $q = 2$ (since each is a 2-simplex and they share a 2-simplex with each other); so being dominating or being aggressive produces the same reaction (3 vertices out of X) for the wife. Also, at $q = 1$, Y5 and Y6 are connected together – so being protective or being gentle produces connected reactions from his wife. If we go to the conjugate complex, KX(Y), we see that the wife's most frequent states are X1 (frightened), X4 (inhibited) and X7 (kind) – so these represent the moods which the wife usually finds herself in. Her attempts to cope with this situation lead to her reactions of X1 (frightened), X3 (bitchy) and X4 (inhibited) being the first group to be connected (presumably being 'bitchy' is her attempt to fight back a bit).

Still in Week 1, we can look at the structure $\tilde{K}$ which represents the husband's reactions to his wife's actions.

In $\tilde{K}$X(Y) – wife's actions/

dimension, q	components of geometry
3	(X1)
2	(X1,X3,X4), (X5,X7), (X6), (X2)
1	(all X's)

In $\tilde{K}$Y(X) – husband's reactions/

dimension, q	components of geometry
3	(Y1)
2	(Y1,Y2,Y4), (Y3,Y6), (Y5)
1	(Y1,Y2,Y4), (Y3,Y5,Y6)
0	(all Y's)

These tell us a compatible story: when the wife is frightened she is responding to four of the husband's moods (most of them, in fact). The groupings of X1,X3,X4 in one piece and X5, X7 in another makes reasonable sense. Notice too in the conjugate complex that the husband is usually dominating (since Y1 has the greatest dimension of 3).

Week 17:
In KY(X) – husband's actions/

dimension, q	components of geometry
2	(Y2), (Y3,Y5), (Y6)
1	(Y1), (Y2), (Y3 . . . Y7)
0	(all Y's)

In KX(Y) – wife's reactions/

dimension, q	components of geometry
4	(X7)
3	(X7), (X4)
2	(X2,X5,X7), (X4)
1	(X2,X4,X5,X7)
0	(all X's)

Although the husband is still aggressive (Y2) he now has equally high simplexes in Y3 and Y5 (protective and affectionate) and in Y6 (gentle). In the conjugate complex, for the wife's reactions, we find that X7 (kind) dominates (at a new high dimension of 4). So the wife spends more time being kind – in response to five different moods of the husband. After that she is mostly inhibited (X4), but we notice too that being supportive, happy and kind have become one component at q = 2 (when in Week 1 she was largely being frightened, bitchy and inhibited).

The other structure, $\tilde{K}$, tells a similar story at this stage in the treatment. Finally we show the structure in

Week 54:
In KY(X) – husband's actions/

dimension, q	components of geometry
3	(Y3)
2	(Y3,Y4,Y5,Y6), (Y1), (Y2)
1	(all Y's)

In $KX(Y)$ – wife's reactions/

dimension, q	components of geometry
5	(X2)
4	(X2, X7)
3	(X2,X7), (X6)
2	(X2,X5,X6,X7)
1	(X2,X3,X4,X5,X6,X7)
0	(all X's)

Now the husband finds that his mood which produces the widest reaction from his wife is Y3 (affectionate) and that this is most highly connected to his moods Y4, Y5 and Y6, at $q = 2$. His wife's reactions are dominated by X2 (supportive) at a new high dimension of 5, and after that by X7 (kind). After that we find X6 (impulsive) and X5 (happy) and, perhaps the most impressive, the fact that X1 (being frightened) only appears at $q = 0$ (therefore in response to only one of the husband's states).

The only point we need make out of the other structure, $\tilde{K}$, comes from the husband's reactions to the wife's action – that is to say, the complex $\tilde{K}Y(X)$. In that, we find that the states Y2 and Y7 *do not appear* at all (they have connectivities $q = -1$!). So John has steadily improved both his potentially unpleasant states as well as those reactions exhibited by his wife. In Week 54 the exclusion of Y2 and Y7 also shows up as *eccentric values* of infinity.

It is worth noting here that the 'vertices' in the sets X and Y actually refer to *actions*, or *active* states in John or in his wife. This means that they are dynamic in nature and so look more as if they should be *traffic* on some underlying backcloth structure – which would describe the husband-wife relationship. Such a structure would clearly be at a lower hierarchical level than these 'vertices', and some attempt to show how it might be regarded is given in the following section. But in the context of clinical psychology it is natural for the dynamics to be the first concern. What the analysis does not give is the dimensionality of this traffic, on the underlying structure. Such knowledge, if it existed – as it surely will eventually – would provide an added insight into the possibilities of the dynamics (of *behaviour* patterns in husband or wife) of the situation, hopefully linking psychiatric and psychological theories.

The love between John and Julie Doe

In Figure 7 we have surmised how John Doe might be fitted into the geometry of his working life and this is compatible with the hierarchical notions discussed under the idea of the collective Self in Chapter 2. But when he comes home John moves into another structure, with its own peculiar geometry, and there he is a family man, a father and a husband. And in that domestic structure his wife Julie lives with her own structural connections; somewhere in it all there must be found the multidimensional space where John and Julie meet as lovers. If being in love (which later becomes the much more complicated relation of being husband and wife and of being father and mother) is already, as we have seen, an expression of the collective $(N + 1)$-Self, then presumably the description of that experience will be at the levels of N, $(N - 1)$, $(N - 2)$, etc. How can it be described at these levels and in appropriate spaces, E^n?

The words 'I love you' surely express, if nothing else, my personal experience of a powerful sense of *unity* with you. Such a notion is also bound to be of a hierarchical nature – we can contemplate being united at the N-level, or the $(N - 1)$-level, etc. – provided we can make sense of it at any one level. If we can express this idea in terms of our geometrical structures it will certainly not be dependent upon any array of theoretical pre-concepts, such as the 'cause' of the condition or any of the myriad analogies which the human mind continually expresses via its poetry, music and the arts. Nor will our mathematical relations exclude such analogies of pre-concepts. Rather will they provide us with a new multidimensional backcloth which can carry new and relevant high-dimensional traffic – which is what a great deal that is written, spoken or sung about love actually amounts to.

This sense of unity, of being in love, can also be characterized dynamically as a state in which *giving* and *taking* are synonymous; such a characterization is naturally an appeal to the notion of traffic which can move between two structures once they have become a new single structure via the identification of vertices and simplexes. This giving and taking will occur in all aspects of the condition – economic, social, cultural, playful and sexual. When I *take* from you I automatically *give* to you (what you *take* from me); when there is *no sense* in which these two things are distinct then we are in love. Of course, this kind of traffic can exist between people who are not in love in the sense that we normally understand it – as between man and woman – but then it will not contain the very powerful sexual features (here I mean the heterosexual features). There is a great deal of discussion about love in religious movements and these I regard as being essentially of the same nature with regard to the idea of unity, but as being relevant only at certain hierarchical levels which I hope to make clear in the sequel. As far as homosexual love is concerned, I see no special distinction therein unless it is meant to include sexual activities which are a substitute for sexual intercourse, or what Wilhelm Reich called the 'genital embrace'. In that case (which I suppose is what people commonly mean to imply by homosexual practices and which they think people are engaged in when they describe them as being homosexual) I think the profound difference between heterosexual love and homosexual love lies in the biological message which descends from the collective $(N + ?)$-Self and which requires procreation at the $(N + 1)$-level. I believe this changes the quality, the dimensionality, of that love and because of the broad awareness of that difference societies usually find it difficult to tolerate too much overt homosexuality. The passing of laws to make homosexuality 'legal' in no way disguises the deep-seated suspicions which the mass of ordinary people feel about it. It is socially dangerous only if it leads to an attack upon the significance of normal sexuality by its protagonists – a theme we shall return to later.

Let us suppose that, at the $(N + 1)$-level, John and Julie can be separately described by similar sets of expressions of their personalities. These will be expressions of their thoughts and their actions as well as of their appearances, and they will be conveniently listed under broader headings of public, private, and sexual. John and Julie will express their love-relationship by being aware of their sense of unity between such expressive sets of vertices. We know of course that this is a tricky situation; people can easily deceive themselves under the very strong pressure of that biological drive which is fuelled by the $(N + ?)$-Self, and the relationship can change with time. But even then we shall see how that change is mirrored in the possibility of change in the appropriate geometry – the geometry which is produced by the union of their separate structures. But at the moment it is sufficient to assume that these pitfalls have been avoided and that the relationship we are talking about is reasonably secure. What then goes into the set at the N-level and the (N — 1)-level?

By the N-level we shall take that which is normally available, in common speech, for the description of the love-relationship. At this level each partner sees, or speaks of, or is sensitive to the other's outward appearance and self-expression (by way of speech and actions in various contexts). At this N-level clothing matters, hairstyles matter, condition and appearance of the body matter, speech and language matter, interests (both intellectual and physical) matter, and expressions of sexuality (such as caresses, gestures, contact, embrace) matter. The identification between *giving* and *taking* is now the essence of being in love at the N-level. The differences between men and women, at this level, are subtle expressions of the sexual differences at the lower (N — 1)-level, etc. If that were not so we would in fact be talking about the affection between friends rather than the love between lovers. At this level, people speak about tenderness, in thought and deed, about sharing likes and dislikes, about kissing and cuddling, about looking handsome or pretty, about being attractive or sexy.

Because we cannot match vertices in two separate geometries which are not to be found there, this view is compatible with the Jungian idea of the anima in man and the animus in woman as the unconscious parts of the psyche which enable us to make satisfactory sexual matches. That in its turn is compatible with early myths about the fall of Man beginning with the separation of the Female from the Male – and the touching idea that everyone must go through life trying hard to match up again with his/her other half.

At the (N — 1)-level we naturally find a set of elements which is covered by those of the N-level. For example the body, at N-level, is just a collection of (N — 1)-bits of the body. We hardly need to take the body apart into its multitude of scientific pieces (that would take us down into (N — 2) or even (N — 3)) – for who is aware of the lover's kidneys under normal love-relationship activity? But the simple 'pieces' of the body will suffice at the (N — 1)-level – like hair, eyes, ... lips, ... arms, hands, breasts, ... waist, genitals, thighs, ... ankles, feet . . .? In addition to this sort of set we shall naturally need the 'units' of speech and expression – the vocabulary and concepts which are used to form the N-level expressions in thought and speech. This will obviously be a very large list of elements but there is no obvious theoretical obstacle to our including them at this level.

At this level, the sense of unity which defines being in love is powerful in its detail – for it must be generating the geometrical structure (in some suitable high-dimensional space E^n) which has to carry the traffic of love-making. This traffic is probably near to the unconscious at the $(N - 1)$-level – surely it is quite so at the $(N - 2)$-level? – and when that is so it must correspond to biological reflex action. Such action transcends conscious thought and is accompanied by the 'surrender' which the $(N + ?)$-Self is demanding. That surrender is not merely the one which should be present in the orgasm of the genital embrace but also is a surrender of identity in that structure which 'being in love' has created in the multidimensional space E^n. So the $(N - 1)$-love is the foundation on which is built the N-love and the $(N + 1)$-love.

Let the poet speak for that $(N - 1)$-love – through E. E. Cummings in the following.

My Love

> my love
> thy hair is one kingdom
>   the king whereof is darkness
> thy forehead is a flight of flowers
>
> thy head is a quick forest
>   filled with sleeping birds
> thy breasts are swarms of white bees
>   upon the bough of thy body
> thy body to me is April
> in whose armpits is the approach of spring
> thy thighs are white horses yoked to a chariot
>   of kings
> they are the striking of a good minstrel
> between them is always a pleasant song
>
> my love
> thy head is a casket
>   of the cool jewel of thy mind
> the hair of thy head is one warrior
>   innocent of defeat

thy hair upon thy shoulders is an army
  with victory and with trumpets

thy legs are the trees of dreaming
whose fruit is the very eatage of forgetfulness

thy lips are satraps in scarlet
  in whose kiss is the combining of kings
thy wrists
are holy
  which are the keepers of the keys of thy blood
thy feet upon thy ankles are flowers in vases
  of silver

in thy beauty is the dilemma of flutes

  thy eyes are the betrayal
of bells comprehended through incense.

And after this impressive list of $(N-1)$-vertices in the geometries, the following poem by Denise Levertov speaks of the N-level unity which is built on that backcloth. Is the woman more sensitive to the higher level unities in this context, I wonder?

Bedtime

We are a meadow where the bees hum,
mind and body are almost one

as the fire snaps in the stove
and our eyes close,

and mouth to mouth, the covers
pulled over our shoulders,

we drowse as horses drowse afield,
in accord; though the fall cold

surrounds our warm bed, and though
by day we are singular and often lonely.

We can summarize the essence of these notions for the love-relationship between John and Julie, in something like the following form:

Level	John's geometry		Julie's geometry
N + 1	Expressions of personality	λ_1	Expressions of personality
	(private, public and sexual)	⟷	(private, public and sexual)
	↑ α		↑ α
N	Outward appearance (clothes, body stance posture, mobility)	λ_2	Outward appearance (clothes, body stance posture, mobility)
		⟷	
	Speech (expressing interests and attitudes)		Speech (expressing interests and attitudes)
	↑ β		↑ β
N − 1	'parts' of the body	λ_3 ⟷	'parts' of the body
	Vocabulary units		Vocabulary units
	↕		↕
N − 2	Unconscious elements	λ_4 ⟷	Unconscious elements
	↕		↑
N − 3	Deeper unconscious	λ_5 ⟷	Deeper unconscious
	↑ ↓		↑ ↓

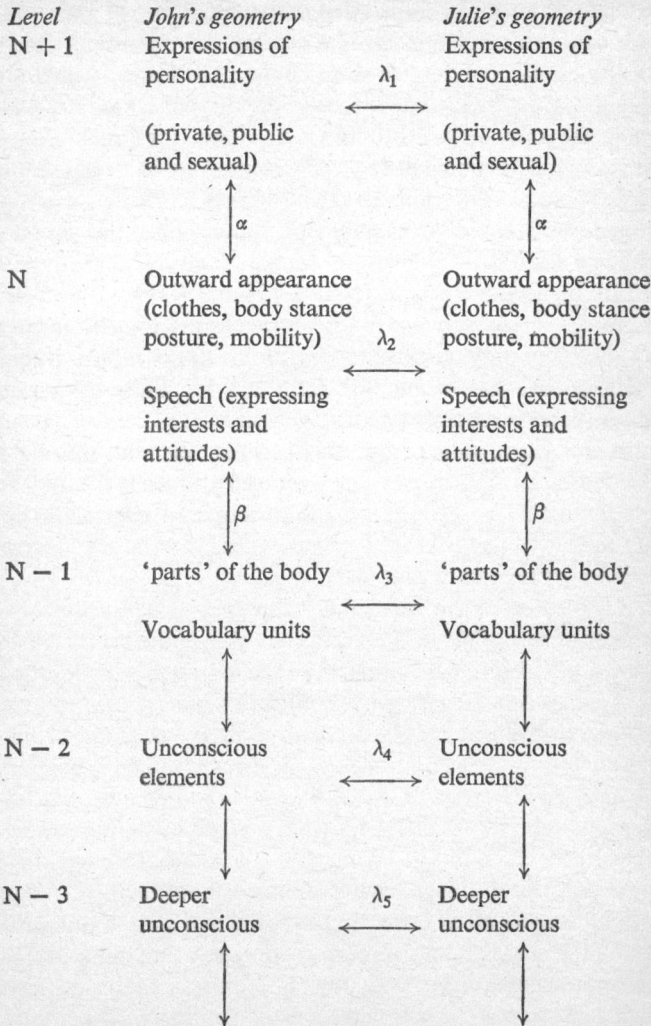

John and Julie will separately possess their own hierarchical structures, which will be multidimensional geometries in suitable spaces, E^n, and these will be complexes at each level. There will be relations like α and β for each of them, and other relations which take us down below the level of consciousness. But whereas John's conscious structures will be what we normally think of as masculine, those of Julie will be what we call feminine. But then, accepting the essentially Jungian view of the human psyche, we would expect the lower levels (in the unconscious structures – that is to say, the not-conscious structures) to contain the anima (for John) and the animus (for Julie). Then the love-relationship between them is represented by the relations like λ_1, λ_2 etc. These relations join up the structures to create the sense of unity already discussed. If elements of, for instance, the anima are contained at the $(N - 2)$-level, in John's structure, then these are covered by higher-level sets at $(N - 1)$, N and $(N + 1)$. But, being unconscious, these covers are not perceived in the N-Self, nor in the $(N + 1)$-Self, but only unconsciously (intuitively) in Julie's structure. So falling in love has meant, for John and for Julie, the intuitive recognition of their separate unconscious Selves; the Beloved becomes the conscious expression of the unconscious Self and, uniting this with the conscious Self, creates a new geometry which is the fulfilment of the individual. That geometry is the structure appropriate to being the new collective $(N + 1)$-Self, and the ancient *content* of marriage has been an expression of that fact.

We notice too that this interpretation means that, since normally the number of sets of elements (vertices in the geometry) increases as we descend the hierarchical levels, then the *largest dimensional spaces* will be those associated with the *unconscious structures* of the psyche. So if Jung is right and many of our dreams are messages from the unconscious, then *the weird world of our dreams will be an immediate experience of what it is like to live and move in a multidimensional space*, E^n. This would be compatible with what mathematicians know about the formal properties of spaces like E^n – the chief of these being the ability to see, for instance, many 3-dimensional faces of a multi-dimensional object all at once. So the scene will shift about without any 'ordinary' sense of linear ordering – the sense that this aspect must precede that aspect (this is only our old 3-dimensional-space experience which is the trap for our conscious perceptions). So perhaps the best way of experiencing our multidimensional geometries is to go to sleep – perchance to dream.

4. Structure and the Creative Man

Whenever we watch colour television programmes we are seeing the results of research work on primary hues carried out by James Clerk Maxwell, back in the 1850s. That work revived the notion of primary colours (and the idea that only three primary receptors were required for colour vision) first advanced by Thomas Young in 1807. Maxwell showed that our visual sensitivity to all the colours of the spectrum can be accounted for by our response to various mixtures of the primary hues of red, green and blue. If the eye receives a mixture (at suitable intensities) of red rays, green rays and blue rays, then the brain interprets the resultant colour as white. (Notice that we are speaking of rays, or hues, and not pigments. Artists have known for centuries that only three primary pigments are needed and that these are red, yellow and blue. But the colours of pigment mixtures are the result of the reflection of unabsorbed rays and so a mixture of yellow pigment and blue pigment, illuminated by white light, reflects the remainder of that white light after the pigment particles have absorbed all they can. Since each pigment actually reflects quite a broad band of colours, around the yellow or blue, it so happens in this case that the resulting reflected light is green.) The colours on the television screen are mixtures of rays (of red, green and blue light) which the human eye sees appropriately.

The point of this is that we have here a simple illustration of the perception of a low-dimensional simplex (a 2-simplex) and its faces. Its representation by a 'triangle' in the geometry appropriate to that structure is possible in a piece of E^3 (actually in E^2, the plane of the paper). It is shown in Figure 8 below.

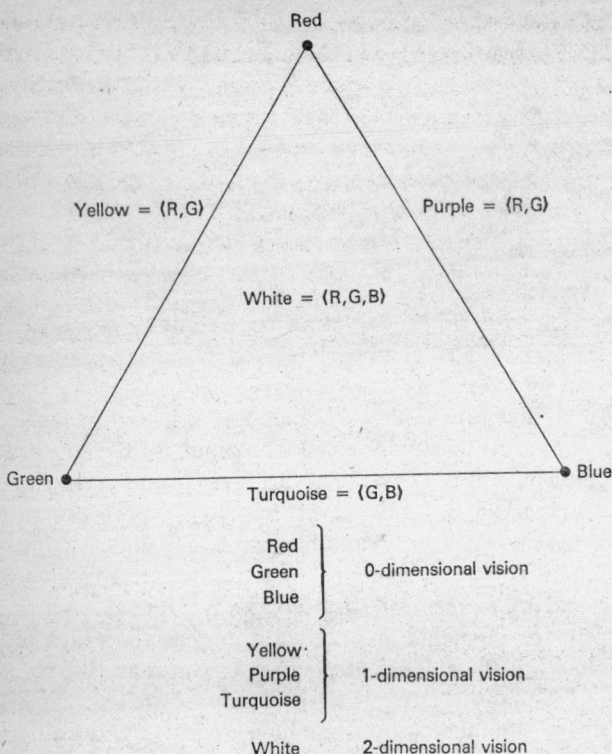

Fig. 8   A 2-simplex of colours in E^2

In this complex we have the following identification:

Red = ⟨Red⟩, Blue = ⟨Blue⟩, Green = ⟨Green⟩: 0-simplexes
    Yellow = ⟨Green, Red⟩, Purple = ⟨Red, Blue⟩
    Turquoise = ⟨Green, Blue⟩: 1-simplexes
    White = ⟨Green, Red, Blue⟩: 2-simplex

So the 2-simplex, together with all its faces, forms a simplicial complex KY(X), where X is the vertex set {Green, Red, Blue} and Y is the set of seven perceived colours. To be able to see all these colours, my vision needs to have the ability to function in the 2-dimensional triangle (in order to see White), that is to say, my colour-vision function must be 2-dimensional traffic on the geometry of Figure 8.

That traffic can move about over the geometry (over the triangle and its edges and vertices) because the structure is suitably connected. Thus when I see Blue that visual traffic is resting on the vertex Blue, when I see Yellow it is resting on the edge joining Green and Red, and so on. My colour-vision function (my colour awareness) must match the dimensions of the geometrical structure which carries it as traffic. For if it does not then I cannot see all the colours which the geometry claims to represent. This gives us an insight into the significance of the dimensionality of the traffic – by noticing the experience when the traffic and geometry might not match (dimensionally).

Suppose my colour-vision is only 1-dimensional traffic. Then I cannot see White, although I can see all the other colours. This is because the traffic (of 'seeing') is then restricted to the perimeter (the edges and the vertices) of the triangle. If we go further and suppose that my colour-vision is only 0-dimensional traffic on the structure, then it is restricted to the vertices only and I can see only Red, or Green, or Blue, and never more than one at a time. If I am shown Yellow then I have to choose between calling it Red or Green; I cannot have both. Maybe I would say it is either and that I cannot say which. Perhaps I would have to resort to our old cowardly standby and say that 'the chance of its being Red equals the chance of its being Green, i.e., 1/2'? If I am shown White I would be faced with the dilemma of having to choose between Red, Green or Blue. So the mismatch between the dimensionality of the traffic and that of the complex can be understood as a problem of having to choose different pieces of the geometry. It is as if in one condition I can see the complex as a hollow triangle, whilst in another condition I can only see it as three distinct points (the vertices) which are not connected. In the first condition the traffic 'sees' the geometry at a 1-dimensional level (it falls into three components at the level of $q = 1$ and into one component at the level of $q = 0$) and in the other condition it sees it only at the 0-dimensional level (now the geometry falls into three components at the level of $q = 0$).

But if my colour-vision function is 3-dimensional (and is therefore greater than the largest dimension to be found in the complex $KY(X)$) then I can see all the colours already described and some more colours not described. In other words I would be aware of the need for another vertex, another primary colour, say Z, to make the triangle into a tetrahedron – so that the 2-simplex of Figure 8 becomes only a face of the new structure. I would be aware of my vision function being cramped, being restricted, being belittled, if it is only allowed to exist as traffic in Figure 8. 'Where are the other colours?' I would say – because there would be the haunting expectation of a super-White (represented by the whole tetrahedron ⟨Green, Red, Blue, Z⟩) and three more edge colours, ⟨Green, Z⟩, ⟨Red, Z⟩, ⟨Blue, Z⟩, and three more triangle colours, ⟨Green, Red, Z⟩, ⟨Red, Blue, Z⟩, and ⟨Green, Blue, Z⟩. How colour-poor the world would be for me if that were the case!

Can this give us insight into the increase in colour awareness which has been described by Aldous Huxley [17] consequent upon taking the drug mescaline? Can such a drug be increasing the dimensionality of the colour-vision function? Does the scientific theory of Maxwell, which produces Figure 8 out of observation by 2-dimensional colour-vision, only reflect the dimensions put into the data by the scientist himself? What sort of complex would we have if Huxley, under the influence of mescaline, had conducted the research into primary hues? Would it be a 3-simplex or a 10-simplex and, if so, would it be nearer to the 'truth' or further away? Is the technological 'truth' we have produced so far only the common experience of lowest dimension, the one we can agree to call 'scientific'? Or was Bishop Berkeley right after all, in a particularly subtle way, and the objective universe is really a projection of our own internal structures?

But then does that leave the artist as the enlightened man who is alone searching at the frontiers of our multidimensional world? If so, it is hardly surprising if his experience often turns out to be such a lonely one. But if we include in the word 'artist' anyone who is in fact searching at these multidimensional frontiers, then it must surely contain not only the conventionally understood artist, who collects his materials (the hardware of his activities) and works them into an expression of some deeply felt truth of human awareness, but also the creative 'scientist' whose hardware might range from bits of rocks to collections of sophisticated concepts (constructs derived from deeper relations between ideas and phenomena). Such a combination of artist, poet, musician, scientist must then be our simple *creative man*; what we need is an understanding of that creative activity and how it can lead to these new frontiers of multidimensionality.

The building of a Master Builder

Our new general kind of artist will need to fabricate and to contemplate various artefacts (utility-oriented constructs) as part of his stock-in-trade, and this will be learned and mastered via an arduous apprenticeship. Such artefacts will range from implements and tools to the preparation of canvases and paints, musical instruments, up to technological things like galvanometers and cathode-ray tubes. With these artefacts men can build more sophisticated structures – the 'middle-art' of paintings, buildings, musical compositions, sculptures, and scientific theories (within a well-practised paradigm, à la Kuhn[9]). Only after that stage can we find the master, the artist/scientist who can create the superstructure by rising above the mere mechanical competence of the middle-art. Now the artist creates a new geometry which transcends that of his contemporaries; his symphony finds new and larger-dimensional spaces which beckon the human spirit, his scientific theory explodes the mental horizons of his struggling colleagues, his painting opens new windows for the soul, and his poetry ricochets our minds around the walls of newly constructed multidimensional palaces.

All this can be described in a hierarchy of sets and relations which naturally give rise to dependent complexes – each of which identifies some geometrical structure in a multidimensional space. In order to achieve this we soon realize that the minimum span of hierarchical levels which will serve our purposes is a set of four; let us call them $(N - 1)$, N, $(N + 1)$ and $(N + 2)$. These are identified in Figure 9, and that diagram has been constructed to illustrate the following ideas.

At the $(N - 1)$-level we place the sets which contain the basic elements for the 'art' we wish to study. We regard these as being of two kinds, the one called *material* elements (and denoted by M_1) and the other called *design* elements (and denoted by D_1) – and of course we could equally well refer to these as $(N - 1)$-materials or as $(N - 1)$-designs. The sets M_1 and D_1 will consist of appropriate things like the following:

Set M_1: pigments, stones, metals, canvas, wood, glass, paper, basic technological things (like electrical conductors, insulators, chemicals etc.)

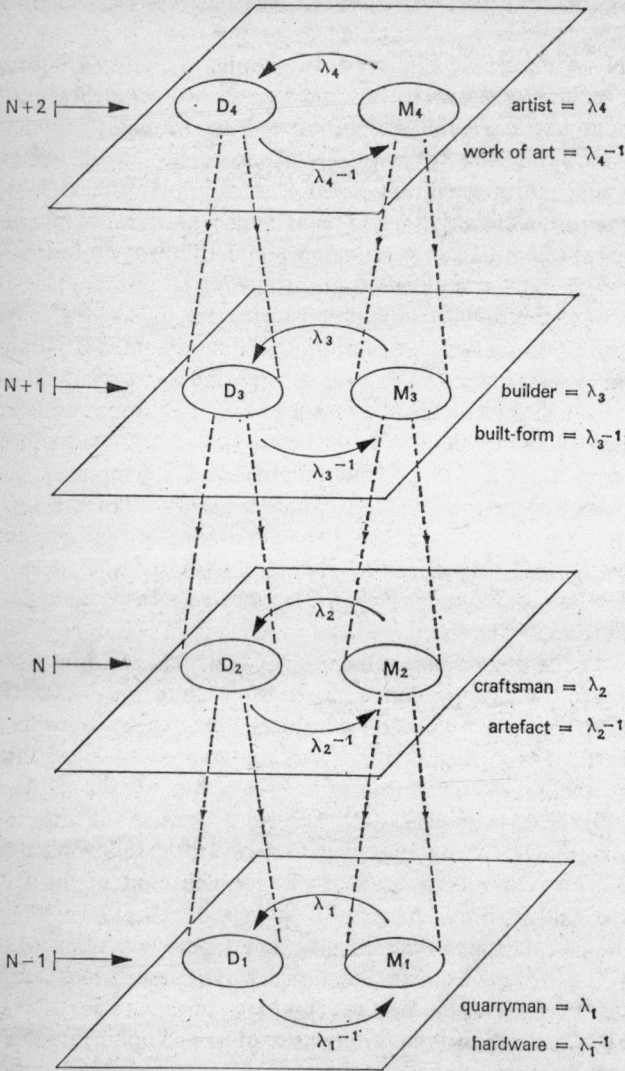

Fig. 9   A hierarchy for the creative artist

Set D_1: various kinds of marks (for specifying points, lines, shapes – on instruments, vessels etc.)

At this $(N - 1)$-level we shall find the supplier of hardware, the quarryman, the manufacturer of bits and pieces, the man or woman who makes the most elementary objects like sheets of paper, lengths of wire, lumps of marble, tubs of clay, pens and pencils, shaped pieces of wood, cans of paint, etc. But how are these things produced? By setting up a scheme of work which takes various members out of the set M_1, attaching various members of the set D_1 in some deliberate manner, and then making that association permanent. This is exactly the process of defining a relation (call it λ_1) between D_1 and M_1 and representing it by a matrix of 0's and 1's, with the members of D_1 as the rows and those of M_1 as the columns. Of course there will be a different relation, like λ_1, for each operator and because of this we can begin to identify the operator by that specific relation. So the relation is the mathematical representation of the operator; more precisely, *the operator* (the 'quarryman') *is represented by the simplicial complex structure*, $KD_1(M_1)$, which the relation λ_1 defines. This is because 'he is known by his works' and a typical 'work' is the geometrical connectivity which links his marks (elements taken out of the set D_1) via the material elements (taken out of the set M_1) which they share. The quarryman, to use the word in a very wide sense, makes a square canvas by fixing four marks (out of D_1) to a thing (a piece of canvas) taken out of the set M_1; he makes it permanent by cutting along the lines – and hey presto! But when the quarryman has produced the hardware he has produced one or more things out of the set M_1 which are connected via the marks (out of the set D_1) they share. This means that *the hardware is the conjugate complex*, $KM_1(D_1)$, defined by the inverse relation $\lambda_1{}^{-1}$. This is how we have indicated the significance of the complexes in Figure 9.

So when the quarryman is a-quarrying he is creating a simplicial complex, $KD_1(M_1)$, via some relation λ_1 which is peculiar to himself. He is therefore representable by that complex, which in turn is representable by a collection of multidimensional convex polyhedra in a suitable space E^{d_1} (we can specify the number d_1 only by precisely laying out the matrix of 0's and 1's which represents the specific λ_1). The hardware he has made is now the conjugate complex, $KM_1(D_1)$, which he has thereby created. This hardware is therefore representable by another collection of multidimensional convex polyhedra in a suitable space E^{m_1}. Notice that, in E^{d_1}, the points are members of the set M^1 whilst in E^{m_1} the points are members of the set D^1. A typical piece of $(N - 1)$-level hardware is a collection of physical things fixed in some precise relation with a set of marks (design elements out of D_1). This pencil (with which the manuscript of this book is being written) is a complex, $KM_1(D_1)$, in which there are five simplices. Their names are taken out of M_1 and are 'graphite', 'wood', 'glue', 'blue paint', and 'gold paint'. Their shapes are defined by elements out of D_1 and they share various common faces which hold them together in one piece. Thus the marks (lines) which define the shapes of the gold letters on the pencil are those very lines out of D_1 which are shared by 'blue paint' and 'gold paint' – since they 'separate' the letters from the blue background (but *we* notice that they are really something *shared* by the letters and the blue background). Thus the pencil is an object in our familiar E^3 – but also a simplicial complex in E^{m_1}, where m_1 is about 9 or 10!

At the next stage in the hierarchy, the N-level in Figure 9, we find a cover D_2 of the set D_1, as well as a set M_2 which covers M_1. The members of D_2 will be subsets of the set D_1 and we shall expect to find associated with this D_2 those people who need and embody the skills of some special selection of the lower level of quarrymen. Such people we shall call the *craftsmen*, for at this level they will be the people who can draw, paint, sculpt, play musical instruments, do carpentry and woodwork, assemble pieces of scientific equipment. The famous scientist Rutherford was a craftsman (an N-level designer) when he constructed his own electroscope out of tin, gold leaf and bits of sealing wax. Beethoven was an N-level designer when he *played* the opening bars of one of his own piano compositions (but not when he composed them). Picasso was a craftsman (an N-level designer) when he actually applied the paint to a corner of his canvas in some specific pattern (which expressed albeit a much higher-level concept devised by Picasso the artist). In this sense the N-level designer is the conjunction of D_2 and a relation λ_2 between it and the set M_2 but, what is more, the members used out of D_2 are those which already possess some organization via the lower relations λ_1, and the members which are used out of M_2 must also be collections of things already quarried (things represented by the structure of some λ_1^{-1}). Thus we really need to regard the set D_2 as a collection of (a cover of) the complexes, $KD_1(M_1)$, whilst the set M_2 is to be a cover of the complexes, $KM_1(D_1)$. This is why, in Figure 9, we must read D_2 as a cover of the set of things D_1-cum-$\{\lambda_1\}$, whilst M_2 becomes a cover of the set of things M_1-cum-$\{\lambda_1{}^{-1}\}$.

It follows that the activity of being an N-level designer expresses a new relation, say λ_2, between these sets D_1-cum-$\{\lambda_1\}$ and M_1-cum-$\{\lambda_1{}^{-1}\}$. Then with this understanding we see that the notion of 'craftsman' is represented by some simplicial complex, $KD_2(M_2)$, induced by a typical relation λ_2. In the same way as before, the dualism between conjugate complexes allows us to see $KM_2(D_2)$ as the *artefact* made by the craftsman. Here we are equating 'craftsman' with N-level designer and 'artefact' with N-level hardware. So the artefact is a complex of M_2-members which is an appropriate structure in a space, E^{m_2}, whose points are members of D_2 – and m_2 is appropriate dimensional number. In the dual sense each craftsman is a structure in some space, E^{d_2}, whose points are members of M_2, and again d_2 is dimensional number.

At the (N + 1)-level we introduce sets D_3 and M_3. The first is a cover of D_2-cum-$\{\lambda_2\}$, so a D_3-member represents one or more of the members of D_2 seen as complexes (craftsmen) in their own right, whilst the second is a cover of M_2-cum-$\{\lambda_2{}^{-1}\}$, and each member of M_3 is a collection of pieces of N-level hardware (artefacts). This corresponds to the fact that an (N + 1)-level designer must be someone who understands (and can therefore use the talents of) a number of N-level designers, whilst the thing he eventually designs is made up of things which have already been designed at the N-level.

Then the (N + 1)-level designer must be the expression of some relation, λ_3, between these sets D_3 and M_3. We have called this the level of the *builder* (the λ_3) whilst the thing he builds (the $\lambda_3{}^{-1}$) becomes the *built-form*. Thus a builder might well be the person who builds a house, then the built-form is that house. He might also be that Picasso who builds a complete picture, and the picture is the built-form. He might be that Beethoven who constructs a symphony by playing and composing, at the technical (N + 1)-level. He might be that Rutherford who constructs a scientific theory of the atom by building his electroscope, using it, and linking it to concepts derived from other experiments. In these cases the picture, the symphony, the atomic theory, are all built-forms. The builders are examples of complexes, $KD_3(M_3)$, in some multidimensional space E^{d_3}, whilst the built-forms are the conjugate structures, $KM_3(D_3)$, in some space E^{m_3}.

Finally we reserve the word *artist* (or *creative man*) for the structure, $KD_4(M_4)$, expressing a relation λ_4 between D_4 and M_4 – themselves covers of D_3-cum-$\{\lambda_3\}$ and M_3-cum-$\{\lambda_3{}^{-1}\}$ respectively. The *work of art* becomes the structure, $KM_4(D_4)$, defined by the inverse relation $\lambda_4{}^{-1}$. The $(N + 2)$-level designer is the artist (the Master Builder) whilst the work of art is the $(N + 2)$-level hardware. The first is a structure in some space E^{d_4} whilst the second is in E^{m_4}, for some suitable numbers d_4 and m_4.

The artist is a multidimensional man, the work of art is a multidimensional thing. In each case the dimensions are high, for clearly we can see from the construction of the hierarchy that d_4 is at least equal to the sum of the numbers d_1, d_2, and d_3. Similarly we expect that m_4 is not less than the number $m_1 + m_2 + m_3$. In either case the d's and m's are characteristic of the field of the creative activity and of the individuals who become the artists.

The artist differs from the builder as the architect (who is such an artist) differs from the builder, as the scientist differs from the technologist, and as the poet differs from the rhymester. The artist, at the $(N + 2)$-level, sees the connectivities between all possible (relevant) builders, so his message transcends that of any one builder – he lives and creates in a higher-dimensional space, which includes the spaces of all builders.

The work of art differs from the built-form as the masterpiece differs from the copy, as the original from the forgery, as the poem differs from doggerel, as the scientific invention differs from the technological development, as the discovery of the universal law of gravitation differs from the invention of a better mousetrap. The work of art exists in a higher-dimensional space than does the built-form; it possesses *content* where the latter only exhibits *form*.

But mankind is not partitioned by the hierarchy of Figure 9. A man can be all things or none. Michelangelo cut his own marble, at the $(N - 1)$-level of the quarryman, and being his own craftsman (N-level) and builder $((N + 1)$-level) he manifests himself as artist $((N + 2)$-level) in impressive works of art. Some will aspire to the

(N + 2)-level whilst being limited by circumstances to the (N + 1)-level, some will be condemned to the N-level whilst their souls cry out for expression at the (N + 2)-level. When Beethoven went deaf he ceased to be an N-level craftsman but could not thereby be denied the role of (N + 2)-artist. Some, of course, will seek the form of being an (N + 2)-artist whilst their talents are appropriate to (N + 1) or even N. Often academic courses and certificates encourage this form of deception. But there are no rules about it, only structure, and it is very easy to get lost and to flounder in a multidimensional space.

The artist at work

This analysis shows us that what an artist *does*, as distinct from what he is, expresses the age-old realization that he is not so concerned with things as he is with *relations*. We can begin to appreciate the problem facing him as he tries to create that work of art, how to give expression to that multidimensional structure which he inwardly perceives. It must be particularly acute for the artist-painter or artist-sculptor because of the low-dimensional space which is his working arena. Whereas the poet spins his structural net in the space whose points are words and concepts, or the scientist builds his structure in a space of mathematical symbols and ideas, and the musician creates his structure in a space of sounds and rhythms, the artist-painter must somehow project the multidimensional geometry down onto a two- or three-dimensional space.

The structure which is apparently there in a naturalistic or realist painting is actually supplied by the viewer, just as he sees the structure in 'real life'. If the artist is successful he no doubt conveys a particular aspect of the multidimensional geometry to the viewer; the famous smile of the Mona Lisa expresses some subtle relation (in some space E^n) which the viewer appreciates, as well as (maybe) the painter. It appeals to sets of vertices to be found in one's experience of human female faces and the cover sets which, higher up the hierarchy, express cultural and philosophical concepts. But the structure of that relation is *not explicitly* represented on the canvas. If it were we would probably not all recognize it, chiefly because it could not wholly be so represented and the bits which appeared might not fit the bits we ourselves could 'see'. In a similar way we must see that Michelangelo's statue 'David' is a projection into E^3 of the sculptor's sense of his (inner) structure in some E^n. This sense must surely then be 'larger than life' since it is to express deep relations which characterize the cultural and psychological associations of the man-hero. The shapes in stone and their mutual relations give the viewer an opportunity to sense the 'bigger space' in which this 'man' exists. In other words it evokes, as with the Mona Lisa, a multidimensional structure out of a 3-dimensional work of art.

I well remember being impressed by a knowledgeable Florentine guide who conducted a party of us around the Renaissance galleries and, summing up his view of that great artistic awakening, remarked that 'the great artists of that time discovered and then expressed the new idea that man is handsome' – and then, giving us all an intense look and quickly correcting himself (in case we were stupid enough to draw the wrong conclusions from his words), he added, 'I mean, of course, handsome in the mind!' In retrospect I would now like to think that he was trying to say that the artists expressed a sense of the multidimensionality of 'man'.

The 3-dimensional man is a mere object, fixed and limited by being trapped in one hierarchical level and regarded as a piece of machinery, full of gears and oils. If technology takes that view of man, the mechanistic view, then little wonder that the human race is fearful of this scientific age. If scientific thinking can do no better than reduce us to analogues of motor cars, then the gulf between science and art, as it is generally understood, can never be narrowed. If we succumb to this mechanistic view of ourselves we must inevitably deny the reality of the sort of structures this book is about, and that will mean denying ourselves. The current Women's Liberation movement in western societies has, as one of its complaints, that 'men treat women as sexual objects'. In so far as this is often true, as I have no doubt it often is, this analysis would suggest that it is a consequence of a much deeper malaise in the technological view of all human beings as 3-dimensional objects. If the dimensionality of a structure is successively reduced it can only be done by destroying vertices in the geometry, as we would expect that the ones which will be left until the last must be those which have the greatest 'power' in the structure. In this case it will be the vertices which carry the traffic of the sexual drives – because these are dominated by the highest $(N + ?)$-Self of the collective species. So these poor poverty-stricken males will spiral down to a view of woman which leaves her only the low-dimensional structure appropriate to these drives. If men are more prone to this than are women then is it to women we will have to look for the rescue? If the 'free woman' tries to conquer this by aping the male, what is left to us all?

Surely the only weapon which can be used to fight this problem is the rational one of trying to develop an awareness of, and sensitivity to, the multi-level hierarchy which contains the multidimensionality of both man and woman. At the mechanistic level there seems to be no defence against having an essentially callous view of man, a view which drives down into the unconscious perceptions of structure which should remain in the conscious – not necessarily as bits of scientific jargon, but more likely and often as strong senses of intuition and of attraction. A sense of beauty about 'man' (who is 'handsome in the mind') can only come with an awareness of his geometry in E^n – the geometry of Self; for beauty attracts and mirrors the Self – and we are back with the love-relationship between John and Julie Doe. So in a very profound sense we must admit to the view that 'beauty is in the eye of the beholder', and even go further than this – beauty *is* the beholder. Seeing beauty means seeing Self – not narcissism, which amounts to an ability only to mirror Self in Self – discovering the geometry in relations found in the outside world which fits the geometry of Self. If one cannot contemplate Self with equanimity, as all religions seem to recommend, then one sees nothing which is beautiful. Then the seeing of the ugly must correspond to the rejection of all observed structure in the thing, and how can one truly reject without destroying, without violence? We shall see later on that this violence can be a struggle with the anti-Self, with which the ugly can be identified.

But to return to our artist, who now carries some of our hopes about fighting the 3-dimensional vision of the mechanist, we find that he is expressing more and more in these times an awareness of the struggle. He has felt the need to find that multidimensional geometry, the more so as technology, posing as science, reached a climax at the turn of the century and just before the holocaust of the Great War – where the 'men as objects' cult reached its insane expression in the hands of the military generals. At that time we find the attempt to put down explicitly on canvas what geometry lies behind (for example) the smile of the Mona Lisa – and I am referring to *cubism* in art.

In this a view of a structure (for example, that of Figure 7 above) is a sort of 2-dimensional tunnel which bores through the multidimensional geometry, collecting up all the faces of the many polyhedra it encounters. Furthermore, these faces will be 2-dimensional or less. They will appear in some appropriate jumble on the canvas. In Figure 10 there is such a jumble taken from some imaginary relation (or relations) defining a person. It is suggestive of a work of art but it is not such a one – but does it suggest someone, two-faced and fork-tongued?

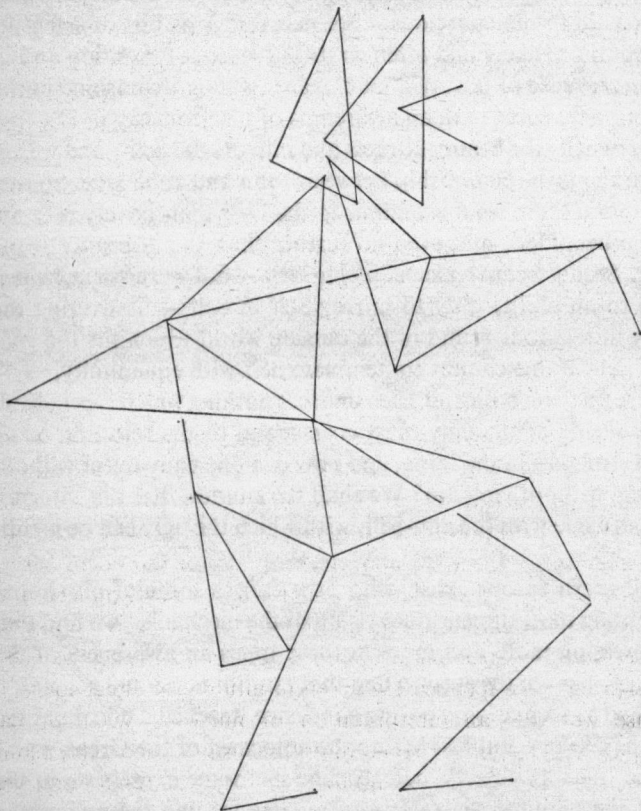

Fig. 10   Who does this remind you of?

On the other hand, the works by Picasso and Braque are serious attempts by those artists to depict the *structures* of their subjects. How reasonable of them to regard these as paintings getting closer to the truth, to the beauty, to the structure of their sitters? In his painting of a seated nude Picasso is saying that the *woman's beauty is a multi-dimensional experience*; the cubist version is a 2-dimensional tunnel-view of that structure. It must have been difficult to understand that in the year 1910 – but notice that about the same time the artist-scientist Einstein was demanding that the world of physics be expressed in a 4-dimensional space (an idea which physicist-technologists to this day find difficult to accept). Picasso's views of M. Vollard as natural-istic (where we have to provide our own intuitive awareness of his structure) and cubist (where Picasso takes a personal look into that E^n) underline this thesis. Also Braque's 'Woman with guitar' leaves us gasping at its daring excursion into a space which is more real than that of ancient Greek geometry (E^3). It is also somewhat amusing to turn these cubist pictures around – which way up should they be? The answer is that 'up' and 'down' are irrelevant concepts in E^n. Only in our familiar E^3, with its gravitational field fixing the vertical, does this question have any meaning. Hang your cubist paintings at any angle, each is equal to any other.

Structure in poetry

The richness of language lies in the many hierarchical levels at which words can be used (with the resulting overlaps which that ensures) and enduring poems have the same sort of appeal as enduring musical compositions. This appeal is through our intuitive awareness of a higher-dimensional space wherein the structure of the poem entices us. Our sensitivity to that higher geometry, to the new connectivities which are now apparent, is manifest by our *emotions*. What we cannot see with our senses in ordinary 3-dimensional space we needs must feel (a sort of 'super-see') with our intuition in a higher space. In this sense we have an interpretation of much of what we call 'emotional' as a 'sight' of some multidimensional geometry (some structure in an E^n, where n is big) to which we can attach ourselves. Whether it is induced by a painting, a Gothic cathedral, or a poem, we speak of a 'widening of the horizons', an 'uplifting experience', or an 'aesthetic awakening'. This in turn suggests that the structure of Self is often being extended or that it is growing and maturing – leaving behind a short-sighted obsession with the obvious physical sense-based geometry of our schooldays. Poems are about this sort of rapport, which is why they are so often about love – between poet and lover, between nature and poet, between poet and history (which unfolds its own connectivity between multidimensional events), or even between poet and the images of poetic structure itself.

We can look at a much-quoted Shakespearean sonnet to find some inkling of this structure. In so doing we are not trying to intrude into the study of comparative criticism nor into the deeper levels of meaning which might require psychoanalysis. We are taking only two rather obvious hierarchical levels and from that standpoint we can find some of the intrinsic structure of the poem. It is a love poem, of course, between poet and lover.

> Shall I compare thee to a summer's day?
> Thou art more lovely and more temperate:
> Rough winds do shake the darling buds of May,
> And summer's lease hath all too short a date:
> Sometime too hot the eye of heaven shines,
> And often is his gold complexion dimm'd;
> And every fair from fair sometime declines,
> By chance, or nature's changing course, untrimm'd:
> But thy eternal summer shall not fade,
> Nor lose possession of that fair thou owest;
> Nor shall Death brag thou wander'st in his shade,
> When in eternal lines to time thou growest.
>> So long as men can breathe, or eyes can see,
>> So long lives this, and this gives life to thee.

At the N-level we shall take a set which consists of those concepts referred to by the nouns Thee (the beloved), May/summer, Sun, Fair (beauty), Thy-summer (the bloom time of the loved one). At the (N + 1)-level we need a set which can act as a cover of this first set. In fact the poet relates the vertices in the N-level set to general ideas (properties), each of which can be associated with a number of things at the N-level; for example, all those things which are lovely provide us with the general notion of 'lovely'. So we identify the following set of vertices at this (N + 1)-level.

A: 'being lovely' (more or less), as in line-2 or in line-7 (fair from fair sometime declines)

B: 'being temperate' (more or less), as in line-2 or in line-3 (the less temperate nature of rough winds) or in line-5 (the more temperate nature of 'too hot')

C: 'enduring time' (shorter or longer), as in line-4 (the short period of 'summer's lease') or in line-9 (the eternal summer) or in line-12 (the eternal nature of this poem) or in lines-13, 14

D: 'growing/diminishing' (life and not-life), as in line-6 (his gold complexion dimm'd) or in line-7 (. . . from fair sometime declines) or in line-9 (. . . summer shall not fade) or in line-12 (. . . to time thou growest) or in lines-11, 14.

The hierarchical scheme is now the following:

Level	Set
$N + 1$	$\{A,B,C,D\} = $ set X
N	$\{$Thee, May/summer, Sun, Fair, Thy-summer$\}$ $= $ set Y

and the poem certainly exhibits the following relation.

Sonnet	A	B	C	D	(reference lines)
Thee	0	1	1	1	1,2,10,11,12,13,14
May/summer	0	1	1	1	1,2,3,4,
Sun	0	1	0	1	5,6
Fair	0	0	1	1	7,8,10
Thy-summer	1	0	1	1	9,10

Of course there are other relations between the actual lines of the poem (numbered 1 to 14) and either of these sets. These relations would give us a structure for the written-form (the built-form already discussed) at each hierarchical level. Although that would no doubt be of some interest, we leave the illustration of that sort of relation to another context (that of the *staging* of *A Midsummer Night's Dream*, below). Here we are trying to express the relation between the sets of conceptual contents in the poem. By deliberately limiting the numbers of the vertices at the N-level we have produced a lower-dimensional structure, one which can be easily represented by ordinary geometry in E^3. The larger-dimensional geometry which this one actually covers could be obtained by investigating the sets at the $(N - 1)$-

level, because there we would need to list all the vertices which, for instance, are covered by the word 'summer', etc. The resulting simplicial complex could hardly be represented in E³, although there would be little difficulty in representing it in a computerized form.

The simplexes in this relation, KY(X), are the following:
Thee = ⟨A,B,D⟩, a 2-simplex, represented by a triangle
May/summer = ⟨B,C,D⟩, a 2-simplex, represented by a triangle
Sun = ⟨B,D⟩, a 1-simplex, represented by an edge
Fair = ⟨C,D⟩, a 1-simplex, represented by an edge
Thy-summer = ⟨A,C,D⟩, a 2-simplex, represented by a triangle.

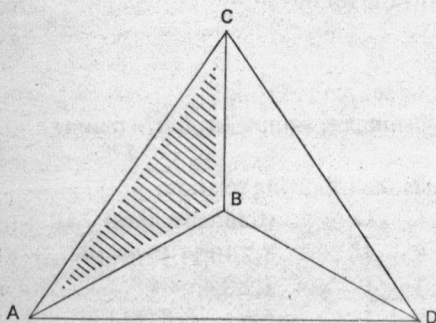

Fig. 11   The geometry of Shakespeare's sonnet

Since the three triangles ABD, ACD and BCD are full (they represent 2-simplexes) the structure is really like a hollow cone with triangular sides meeting in the vertex D – the empty triangle ABC being the mouth of the cone (where the ice-cream goes).

This structure must carry any appropriate traffic, such as that of the reader's attention (in which he recognizes the concepts of the poem and their mutual relation). If he can hold together three concepts simultaneously in his mind then his response to the poem is as 2-dimensional traffic; then he appreciates the poet's Thee in its relation to the triad, A,B,D, and so on. In this sense we can think of the reader as identifying with the whole structure (the surface of our ice-cream cone). But if the reader can only grasp the conjunction of at most two of the concepts (out of A,B,C,D) then his perception is manifest as 1-dimensional traffic on the structure. His attention can then only live on the edges, never on the triangular faces. So now his awareness of the dimensionality of the poem is that much inferior to that of the first reader (who was 2-dimensional traffic). Our 1-dimensional reader is missing a certain *feeling* about the poem, a feeling which the other reader can never transmit to him.

I submit that this makes the point about the significance of the structure's dimensionality and geometrical connections, and that point is quite independent of whether or not Figure 11 is a 'good' representation (at this level) of the poem.

Let us pursue the geometrical implications further, and concentrate on the 2-dimensional reader first, calling him R_2. He can grasp three concepts, out of the set $\{A,B,C,D\}$, at once and he can form a unity out of those three. This is what we mean when we say, rather loosely, that he can live on a 2-simplex, such as Thy-summer which is $\langle A,C,D \rangle$. Thus he can associate with the concept of Thy-summer the triad-as-one whose vertices are A = being more or less lovely, C = enduring (or not) in time, and D = growth/decay (life and death). Naturally, faces of triangles (that is to say, edges and vertices) cause no trouble to R_2. The concepts of Sun and Fair, in their relations to the members of the set X, are merely edges of May/summer and of Thy-summer respectively. To R_2 the poem presents this feeling of a hollow cone with vertex D, so D plays a central role in the poem: R_2 thinks that just as a cone turns on its vertex so this poem rotates about the notion of growth/decay (life and death).

But the reader R_1, whose perceptions allow him to move only on the edges of the structure, cannot see the solid triangles at all. To him the notion of the cone is not only irrelevant but actually incomprehensible. Since all the edges (six precisely) are in the structure there is no obvious key vertex. The role of D is no more significant than is that of A,B or C. The poem is about 'being temperate', 'being lovely', 'enduring in time' and 'growth/decay' – nothing to choose between them. Furthermore because R_1 cannot appreciate, for instance, the simplex Thee he only identifies this 'thee' in the poem as any one of the three edges $\langle AB \rangle$, $\langle BD \rangle$ or $\langle AD \rangle$. So his concepts of the members of the set Y are weaker (of lower dimension) than are those of R_2. Already (and the difference between R_1 and R_2 seems so slight – a single dimension!) R_1's emotional reaction to the poem is more feeble than is R_2's.

R_2 feels the 2-dimensional nature of the structure, it *feels* more connected, more secure, whereas R_1 feels that the structure is full of *holes* – four to be precise, just windows without glass. Of course in a sense R_2 sees a hole (the shaded triangle, where the ice-cream goes in) but it is not a 'real' hole because it is covered over by the rest of the cone. If R_2 goes into his 'hole' then he must come out the same way he went in. Not so with R_1, his holes are real. To him there is just not enough geometrical surface to cover them up. So R_1 *can only go around his holes*, he cannot go across them. He cannot get to $\langle B \rangle$ from $\langle CD \rangle$ without moving his perceiving traffic around the hole whose edges are $\langle CD \rangle$, $\langle DB \rangle$ and $\langle BC \rangle$. In contrast R_2 would have no such difficulty because R_2 can occupy (be conceptually aware of) the whole triangle $\langle BCD \rangle$ and this automatically links the vertex B with the side CD: R_2 can get from B to CD without going anywhere!

But now R_1, having to go around the hole B→C→D→B, *experiences that hole as an object* (obstacle) in the structure. An object is something you have to go around, you cannot sail blithely through it, and that is because there is no path (no geometrical points) through the thing. So *holes in a structure KY(X) are experienced as objects around which traffic is forced to divert*. To a typical R_1 it looks as if the poem is a sort of string bag and he must keep moving around all the holes/objects in order to get the 'feel' of the thing. Of course, by 'he' we understand the traffic of perceptive attention which he contributes to reading and appreciating the poem – the semantics of it.

So if the poem contains a much more complicated structure (as indeed does this Shakespeare sonnet if we go down to the $(N-1)$-level sets) in a higher-dimensional space, E^n, the appreciative reader needs to be some suitable R_n (the n's must match). If not, it is highly likely that the poem will seem to contain at least one (maybe many) high-dimensional hole(s). Getting the 'feel of it' will be a continuous run-around, bouncing off these weird holes/objects.

The holes which R_1 has noticed in Figure 11 should be called zero-dimensional holes (written perhaps as 0-holes or 0-objects) because they are formed by edges which are zero-connected (through the 0-simplexes at their ends). There will then more generally be things like 1-holes (or 1-objects), 2-holes etc. These will be more difficult to sense but will equally well represent obstacles (to the free movement through the geometry) of 1-traffic or to 2-traffic, etc. For example, a 1-hole could be formed as a sort of hollow 'cylinder' whose sides are made of triangles which meet in edges. Such a thing is illustrated in Figure 12, where the 1-hole is indicated by the sequence of triangles

$$1 \rightarrow 2 \rightarrow 3 \rightarrow 4 \rightarrow 5 \rightarrow 6 \rightarrow 7 \rightarrow 8 \rightarrow 1$$

So if q represents, once again, any of the natural numbers, $0, 1, 2 \ldots$ we might naturally speak of a poem, or a painting, or a man, containing so many q-holes in its structure. Then these will act as a new sort of run-around object for any appropriate q-dimensional traffic which is trying to live on the structure.

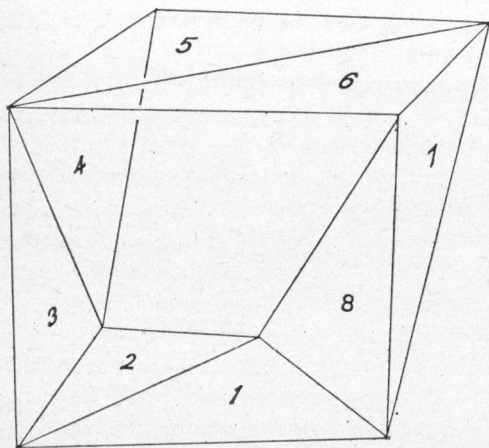

Fig. 12   An example of a 1-hole

Punching q-holes into the geometry

At the $(N + 1)$-level set of data our Shakespearean sonnet acquired a 'hard' structure which we illustrate in Figure 11 (on page 124) and our two readers of that poem, R_1 and R_2, constituted 1- and 2-dimensional traffic (of understanding) on that structure. Whilst R_2 could not 'see' any holes in the structure, R_1 naturally thought that it contained a number of 0-dimensional holes. Now in Figure 12 we have illustrated the connectivities between simplexes which give rise to a 1-dimensional hole (a 1-hole), and we have seen that in that context 1-traffic must 'bounce off' such a 1-hole, making it appear as an actual *object* in the structure (of whatever it is that Figure 12 represents). Our reader R_1 bounces off the 0-holes in Figure 11 because he (or his understanding of the connected concepts in the poem) is constrained to keeping going around them. In fact the structure is so simple (to R_1) because it only consists of edges (like thin rods) joined up at their ends. But what effect, in the general case, do these q-holes have on q-traffic? Is there something we can say about such a situation which is not context-specific?

To give substance to our ideas let us suppose that Figure 12 refers to some part of the structure of John Doe's work environment. We have already treated that environment at the $(N + 1)$-level, as in Figure 7, but now we can suppose that Figure 12 represents part of the N-level geometry of that environment. Let the eight simplexes represent eight offices in the firm which deal with lower-level matters than the original $(N + 1)$-set X (on page 88). In other words the vertices in Figure 12 (there happen to be eight of them also – an accident only) refer to some of the minutiae of running the business, like matters affecting named individuals both inside and outside the firm, or specific details of supplies or transport, or references to specific sales outlets, and so on.

By going round various edges in the 1-hole it is easy to identify various 0-holes as well: so Figure 12 illustrates both dimensional kinds of q-hole in the firm. This means that there will be a great deal of 0-traffic and 1-traffic which can only *circulate* around these holes – bouncing off the objects which they constitute. This happens because there is plenty of traffic (matters of business, like sending messages, letters, memos, and having discussions about making decisions which affect orders, goods, sales, and individuals) which needs to move across from one office to another – since vertices are shared and many matters are legitimately the concern of more than one office. But all this traffic will be at the appropriate level, which we are calling the N-level, and so it will often be subordinate to lots of $(N + 1)$-traffic – this latter being the traffic which is concerned with policy matters, and which is formally in the hands of the six executives (like John Doe) whom we have listed on page 90 as members of the set Y. Our N-level offices are naturally occupied by N-level staff – who will be answerable to the firm's policy-makers and executives. So now we can see that quite a lot of this N-level traffic, which is moving around our 1-hole, is the kind which cannot come to rest without the intervention of some (N + 1)-executive (who effectively makes the 'decisions' – even when some of this is properly delegated to N-Staff). Now imagine that the 1-hole exists in the structure and that, at the same time, there is no (N + 1)-level executive there to bring the circulating traffic to a halt: then it must continue to go around the 1-object in the geometry.

This kind of permanently circulating traffic, which has to keep moving around the q-objects in the structure, I intend to call *noise* in the geometry. In this case it would be noise at the N-level. The word 'noise' is appropriately taken from the field of electronic communications, where it is associated with the 'signal'–to the latter's detriment; too much 'noise' to 'signal' in electronics messes up the communication channels, and destroys the TV picture most effectively. So electronics 'noise' gets in the way of proper and efficient message carrying: so it is with John Doe's working environment when there is too much noise in the structure – getting in the way of efficient decision-making traffic.

If it so happens that it is John Doe himself who sits above this 1-hole, at the (N + 1)-level, and so controls the circulating traffic, then he is the 'office' which is well connected to all those eight simplexes in Figure 12 and, via these connections, he effectively 'fills the 1-hole' (from above). So his important function, as an (N + 1)-executive, is to reduce the noise around the N-level q-holes; that is a measure of his success as an executive. If it happens that John Doe is an amateur 'empire builder', then it is in his interests to create lots of q-holes underneath him (at the N-level) so that his subordinates are trapped in a flow of noise – which their activities inevitably and legitimately make. He achieves this by setting up 'offices' (which are identified as 'jobs which need doing by individuals') at the N-level which are answerable to him alone. Then he sits back and lets the traffic start to circulate! It seems inevitable that the mere presence of a q-hole in a structure produces traffic noise – perhaps this is close to being a 'law of companies/organizations'? It is certainly the bane of institutions in the following sense.

During 1973 and into 1974 I was directing research into institutional and community structure and, in particular, examined the formal committee structure (at various hierarchical levels) of the University of Essex – with the encouraging support of the Vice-Chancellor and the collaboration of the administrative staff. That showed that the structure at $(N + 1)$-level contained a surprisingly high number of q-holes (something like fifteen to twenty, at various values of q): this was in the *formal structure* derived via the relation between the set of committees and the set of academic staff who were sitting on them.[20] What was particularly interesting was the fact that in many cases *there was no formal committee at the* $(N + 2)$-*level which could be regarded as filling in the holes*. So, on the face of it, there was a great deal of traffic noise circulating throughout the geometry – meaning that decisions were difficult to arrive at, in many areas of university business. But since decisions *must* be arrived at sooner or later, in order to keep the thing going, it was obvious that some *informal* committee must be found to fill in the appropriate q-holes. In most cases, of course, that informal committee had to be the Vice-Chancellor himself (who else, for God's sake?) – but this situation naturally generated the *feeling* throughout the community of some 'power behind the scenes' acting to outwit the formal structure. So it is, of course, but the fault lay not with the Vice-Chancellor but rather with the (unwitting) structural weaknesses manifest by way of the q-holes! Nor is this situation going to be peculiar to the University of Essex, for it is clearly an institutional hazard. How many q-holes, with their accompanying inducement to traffic noise, can we find in the organizational structure of Local Government or the Civil Service?

No doubt some noise is inevitable in any large organization, and there are bound to be many 'pseudo-committees' trying to fill holes 'behind the scenes' in an effort to make the organization function, but it is also inevitable that all this traffic is expensive: it costs a great deal of money to keep it flowing. And in that sense the top-heavy organization (which probably only means an organization with too many q-holes in it) is wasteful of resources and manpower. So what contribution to inflation is this kind of inefficiency making throughout our whole society? How can we avoid the mistakes of reorganization (in both public and private administrations) which so warp the geometry that the number of q-holes just keeps getting larger and larger? How can we stop the incestuous growth of administration which grows only in loops and generates only noise – except by seriously studying the geometry of these structures in their multidimensional spaces?

The alternative can be social disaster in many spheres – as Walpole appreciated in his famous remark that 'everything's at sea, except the fleet'.

The built-form of *A Midsummer Night's Dream*

Before we regard a play as a work of art we also need to see it as a built-form, and in that sense it will possess a structure which producer and actors must master before they can hope to transmit any other feelings which the play can exhibit, *qua* work of art. To many audiences it might well be all that the play consists of, but that depends on their receptivity to super structure inherent in the work of art as well as to the effectiveness of the production. To see the built-form in this Shakespeare play we do not need to analyse it as an enduring work of art, but we do need to see how the bricks have been laid together to give us a backcloth which can carry suitable theatrical traffic. To do this we first notice that the play offers us three groups of sets – the latter being related at various hierarchical levels, as follows:

group A: the play, the acts, the scenes, the subscenes

group B: the characters

group C: the commentary, the play, the plots, the subplots, the speeches. (For the moment we leave out 'the commentary', since we shall see that it has an interesting and special role.) These sets of things, within the groupings denoted by A, B and C, may reasonably be fitted into the following hierarchical schema:

Hierarchy H:

Level	Sets of elements
N + 2	the play (as in group A), and also the play (as in group C)
N + 1	the acts (A), the plots (C)
N	the scenes (A), the characters (B), and the subplots (C)
N − 1	the subscenes (A), the speeches (C)

At the (N + 2)-level the sets are particularly simple and contain one member only, 'the play' – that is to say, the name of the play. This may of course be viewed as in group A or in group C. The difference is that, if we view it in group A it is a cover of the acts, scenes and sub-scenes, while if we view it in group C it is a cover of plots, subplots, speeches. But we shall not find much structure at $(N + 2)$, per se – in a set with one member. There is, however, a natural relation between $(N + 2)$ and $(N + 1)$, which simply expresses the fact that 'the play' contains the 'five acts'. Then going on down to the N-level, this gives us the fact that these five acts form a cover (actually a partition) of the scenes, and so on. The same goes for a similar simple set of relations which link the play with the plots, the subplots and the speeches. But all these relations are simple partitions and in fact probably express chiefly the tradition of the editing of the play. After all, it should be noted that Shakespeare did not necessarily write in terms of acts and scenes; the practice of dividing his plays (partitioning them) in this way was adopted by later editors.

It is more important to look at the relations which are built in to the hierarchy H, which reflect the interaction between the characters and the plots etc. And here I shall only illustrate the structures involved by selecting one or two relations. To do this we need to list the elements of the sets we intend to use, as follows:

The plots (at $(N + 1)$):

P1   The court of Theseus, comprising in the main the preparations for his wedding and performance of the play by Bottom and his friends, and the consideration of and resolution by Theseus of the problems surrounding Hermia and her two suitors

P2   The world of the fairies, comprising the quarrel between Oberon and Titania and the trick played by him on her with its consequences for both Titania and Oberon

P3   The world of the lovers, comprising the basic predicament of Hermia, Helena, Demetrius and Lysander and the interference by Oberon and Puck

The subplots (at N-level):

p1   Theseus' role in the lives of the four lovers
p2   The relationship between Theseus and Hippolyta and their wedding celebrations

p3 The rehearsal and performance of the play by Bottom and his friends

p4 The general world of the fairies – their songs and powers

p5 The relationship between Oberon and Titania

p6 Oberon's trick on Titania and its consequences for her and Bottom

p7 The basic predicament of the four lovers

p8 The interference by Oberon and Puck in the lives of the four lovers and its consequences

The characters (at N-level):

Theseus	Oberon	Quince
Hippolyta	Titania	Bottom
Egeus	Puck	Flute
Hermia	Peaseblossom	Snout
Lysander	Cobweb	Starveling
Demetrius	Moth	Snug
Helena	Mustardseed	
Philostrate		

There is an alternative list of characters, which I shall use in response to the Royal Shakespeare Company's production of the play – directed by Peter Brook – in which Theseus and Oberon were combined into one character, whilst Hippolyta and Titania became another single character. When using this alternative list we can refer to it as the Peter Brook set.

[1] First we look at the relation between the set of plots (call this the set $Y = P1, P2, P3$) and the set of characters (call this the set X). This relation gives a typical matrix array of 0's and 1's, and analysis (a computer programme was in fact used to obtain all these analyses) gives the following simplicial complexes.

$KY(X)$ is the structure in which the plots are simplexes whose vertices are characters. In this complex we find that when $q = 7$ (the 7-dimensional level) the plots are disconnected so there are three 7-components, that is P1, P2, and P3. (See the discussion about the geometry of Figure 7 in Chapter 3 for a reminder about the idea of 'component'.) But at the 6-dimensional level (where $q = 6$) we find that P1 and P3 become connected (they share a 6-dimensional face) and we then have two components, (P1, P3) and P2. At the next level, of $q = 5$, all three plots become joined in one single component, (P1, P2, P3) – so then the geometry is all in one piece. What does this mean?

In the first place, we remember that the things which connect the plots are numbers of characters. So at the 7-dimensional level no two of the three plots share any set of eight characters, but at the 6-dimensional level we are told that plot P1 and plot P3 share seven characters although neither shares seven with plot P2, and at the 5-dimensional level any two of the plots share some set of six characters. These connectivities now determine what dimensional traffic can move about through the geometry, and perhaps the most important traffic we should consider is that defined by the intelligent attention of the audience. If the audience is capable of intelligently following the characters eight at a time, then that attention constitutes 7-dimensional traffic (or 7-traffic) on the structure of this complex. In that eventuality the audience will be able to perceive the plots as three distinct entities. On the other hand if the audience constitutes only 6-traffic (being able to follow simultaneously the fortunes of seven characters) then it will see the play as being composed of two pieces – one being the plot P2 and the other being the component (P1, P3). In this case the audience will find it difficult to separate the plots P1 and P3 (the court of Theseus and the world of the lovers). Finally, an audience which is 5-traffic (cannot follow more than 6 characters at once) will think the play is one single but confused plot which involves the three features we call P1, P2, and P3.

We can profitably look at the conjugate structure, KX(Y), in this relation; now we find that the characters (who are now the simplexes, with vertices taken out of the set Y of plots) which dominate are the lovers Hermia, Helena, Demetrius and Lysander. Each of these is the same 2-simplex ⟨P1, P2, P3⟩ whilst each of the other characters is a face of this. An audience which constitutes 2-traffic on this structure is one which can intelligently follow three plots at once. Such an audience sees the play as being primarily about the four lovers, and all the other characters become subordinate to these.

[2] If we look at the parallel relation between the plots Y and the Peter Brook set (call it Z) of characters, then we find a change in the geometry. In the structure KY(Z) the chief difference is that at $q = 6$ all the plots become one component, so now an audience which is 6-traffic sees essentially one plot. In the conjugate structure, KZ(Y), we

find that the dominant characters are now the set of the four lovers together with the two extra characters (joint ones) Theseus/Oberon and Hippolyta/Titania. So this uniting of the characters has elevated each to a new level which is the same as that of the four lovers.

[3] If we look at the relation between the set Y of characters and the set X of subplots $\{p1, \ldots p8\}$ we obtain two complexes again. In $KY(X)$ the structure is dominated by subplot p3 since this turns out to be a 14-simplex. Thus an audience which follows fifteen characters simultaneously sees this dominance; in fact this dominance will persist until we reach the level of $q = 7$, because only then do other subplots enter the structure. The new ones which enter at that level are p2, p5, and p6. These concern the relationship between Theseus and Hippolyta, the relationship between Oberon and Titania, and Oberon's trick played on Titania. These fall into two separate components, (p2, p3) and (p5, p6). An audience which is only 7-traffic will regard these two separate components as dominating the play – and even then it will not easily separate p2 from p3, nor p5 from p6. In fact we can list the computer output for this structure as follows.

$KY(X)$:	q-value	separate components
	14	(p3)
	7	(p3,p2) (p5,p6)
	6	(p3,p2,p1) (p5,p6,p4)
	5	(p3,p2,p1) (p5,p6,p4) (p8)
	4	(p3,p2,p1,p8,p7) (p5,p6,p4)
	1	(p1,p2,p3,p4,p5,p6,p7,p8)

Interestingly we notice that at $q = 5$ there are three separate components, so at this level of 5-traffic an audience sees the subplots falling into three pieces – they think there are only three subplots, not eight. Further down, we see that an audience which can only muster 1-traffic (can only follow the fortunes of two characters at once) will see the play as one single plot – a confusion of our eight subplots.

Turning to the conjugate complex we get the following computer analysis.

KX(Y):	q-value	separate components
	4	(the four lovers) (Oberon) (Puck)
	2	(lovers, Theseus, Hippolyta, Egeus) (Oberon, Titania, Puck, four fairies, and Bottom)
	1	(everyone except Bottom's friends)

Thus an audience following five subplots ($q = 4$) sees the dominant characters as the lovers (as a single group), Oberon, and Puck. If the audience can only manifest its interest as 2-traffic (following three subplots) then the characters form themselves into two large groups which roughly reflect the division into the mortals and the fairies (with Bottom as an extra).

[4] If we go to the Peter Brook set of characters and study the analogous relation we find no significant difference in the structure KY(Z), from that of KY(X). But there is a difference in the conjugate KZ(Y). We find that Theseus/Oberon dominates with a very high top-q of 7, as follows:

KZ(Y):	q-value	separate components
	7	(Theseus/Oberon)
	5	(Theseus/Oberon, Hippolyta/Titania)
	4	(Theseus/Oberon, Hippolyta/Titania, four lovers, Puck)
	2	(all the above, and four fairies, and Bottom)

Eventually, at $q = 0$, all the characters fall into one component. What is striking about this structure is that every character is a face of (a sub-polyhedron of) the dominant character Theseus/Oberon. So now the audience must see the play as dominated by this single (joint) character, no matter how many subplots it can manage (provided that number is greater than five). For lower dimensional traffic, like 4-traffic, the audience will think that there is little to choose between the members of the one component – so it is all about Theseus/Oberon or equally about Hippolyta/Titania, or equally about the four lovers. This constitutes quite a big change from the structure when the conventional list of characters is kept separate.

[5] Finally, we look at the relation between the set of 9 scenes (call this the set Y) and the set X of characters. The scenes are numbered through

the play, two in each of the first four acts and one in the last act. In the structure KY(X) the dominant scene is the 9th, with a top-q value of 19 (includes 20 characters) – this is a mopping-up scene, of course. But the next scene to enter the structure is the 7th at $q = 14$. All the scenes enter the structure only at $q = 5$, so the audience must be able to follow six characters to be aware of all nine scenes, as an entity.

In the structure KX(Y) there are two dominant components at $q = 5$. These are formed by (Demetrius, Helena) and by (Puck). So an audience which can manifest 5-traffic (can comprehend six scenes at once) sees the play as essentially about Demetrius and Helena on the one hand, and Puck on the other. At $q = 3$ the characters fall into two separate components again and the majority have then entered the structure. An audience which follows four scenes at once (constituting 3-traffic on the geometry) sees the play as being about, on the one hand, the lovers, Puck, Oberon, Titania and the fairies, and on the other, Bottom and his friends.

[6] What happens to the above relation when we change over to the Peter Brook set of characters, Z? Well in KY(Z) there is no significant change from the previous KY(X). But in KZ(Y) the structure is now dominated by three components at $q = 5$, as might be expected, and an audience following six scenes sees the play as being about Theseus/ Oberon, Demetrius and Helena (as one piece), Hippolyta/Titania (as another) and separately about Puck. Once again the combining of characters has increased the dimensions of the joint characters, in this relation.

It is interesting to see how speculative changes in scenes and/or in characters induce geometrical changes in the structures of various relations and that this kind of analysis serves to monitor those changes. Going further down into the $(N - 1)$-level, where it is necessary to identify subscenes and the contents of speeches, would naturally give relations which would need large matrix arrays for their representation. But in that case the computer analysis, which sorts out all the connections, would cope with the problem quite well and efficiently. So speculative building of the built-form can be sorted out overnight in the computer and delivered onto the producer's desk next morning. Would Will Shakespeare have found that entertaining, I wonder?

The idea of a meta-hierarchy

The above discussion about the play as built-form deliberately left out 'the commentary' – that which is spoken by Puck at the end of the play. This was because it naturally falls into a separate category, in that Puck makes a direct appeal to the audience in the theatre – as he says,

> 'If we shadows have offended,
> Think but this, and all is mended,
> That you have but slumber'd here,
> While these visions did appear . . .'

But this means that 'the commentary' is outside the play, is one with the audience, or (in terms of the built-form structure) it is *outside* the *hierarchy*, H. Only from that standpoint is it possible to comment on the play, just as the (intelligent) audience sits and quietly comments to itself as it watches the play – or goes home and remarks about it later on. And from such an outside position it is possible to contemplate *altering* the structure of the play, that is to say, altering the relations and sets which define the hierarchy H.

The same situation arises in *The Tempest*, where Prospero speaks an Epilogue:

> 'Now my charms are all o'erthrown,
> And what strength I have's mine own –
> Which is most faint: now 'tis true,
> I must be here confin'd by you
> Or sent to Naples . . .'

In other words the further *action* of the play is thrown open to the intervention, in imagination, of the audience – so the hierarchy H is not an absolute thing; it can be modified and extended by someone who has a mind to it and *who is outside it*.

This outside position, which is being appealed to, can best be described as one of a *meta-hierarchy*, and since it is immediately above H, in some sense, we can naturally denote it by $(H + 1)$. In this meta-hierarchy there will be a set of meta-hierarchical levels (we could call them things like M, $M + 1$, . . .) and each of these will contain meta-relations and meta-sets (the 'meta' is relative to the hierarchy already established, that is, H). The elements of a typical meta-set will be the *relations* which are found in H (or which could be found in such an H) – because then the meta-relation will tell us how one typical H-structure is related to another such H-structure. Changing any H-structure (like changing the built-form of *The Tempest*) will be the expression of such a meta-relation. Naturally the play's producer and actors need to occupy a meta-position in order to present the work, and the audience needs to do so as well in order to appreciate the production. So what is normally called the 'higher criticism' must constitute traffic (which has therefore become *meta-traffic*) on the structure which is defined by possible meta-relations between suitable meta-sets culled from the hierarchy H (where the original action is to be found).

A pictorial representation of the hierarchy H and the meta-hierarchy H + 1 is shown in Figure 13. It is a situation which we commonly find. For example, in government there is a hierarchy, H, which describes the structure of all Local Authorities – what sets they consist

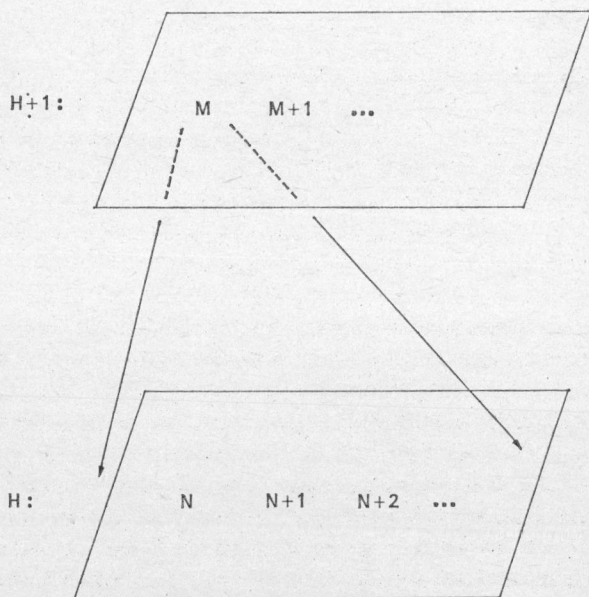

Fig. 13   Hierarchy and meta-hierarchy

of and how they are to interact (via relations between those sets) – the councils and committees and who is to elect them and when. But all this structure is decided by the Central Government, who can make the rules (and so replace them by a new lot of rules). The central government must therefore be in the meta-hierarchy, $H + 1$, relative to its brainchild 'local government'. When there is a reorganization of local government the laws which are passed by the central government constitute meta-traffic in that meta-structure. This function is of course often mixed up with functions which the central government takes upon itself and which are traffic on the lower hierarchy, H. The innocent citizen is someone who sits underneath the whole thing – paying his taxes and his meta-taxes!

The same point is forcefully made by referring to the game of chess – which is typical of many decision-making situations, where tactics and strategy (positional play) intermingle. For, as discussed in Chapter 1, there is a hierarchy, H, of structure at any position during the game. This is defined by the relations between the pieces and the squares of the board (at more than one level), and this relation changes with every move which is made. Thus the hierarchy H changes with every move, and so the Master player must really operate at the meta-hierarchical level *when he is deciding on which move to make*. So then he is probably occupying the level M in the hierarchy, $H + 1$, and the level $M + 1$ in that meta-hierarchy will be the level at which he contemplates *sets* of strategies (this being a cover of strategies at the level M). This must be the level where the personal style of the Master is embedded – because he will have certain preferences in the cover he uses (out of the meta-level $M + 1$).

A similar situation is apparent in the economic life of the nation, where there is a hierarchy, H, made up of the relations (at various hierarchical levels) between prices and commodities. The flow of money in that structure will constitute traffic (on the various complexes) in the geometry, the flow of commodities being traffic on the conjugate parts. But in addition there is a meta-hierarchy in which these relations may be transformed – as the economy ebbs and flows – and where planning the economy is located. Furthermore, there will be meta-traffic (which is really meta-money) which will be qualitatively different from the traffic. Such meta-money will be what is used to trade money with – for only in $H + 1$ can we speak of the 'price' of money (the H traffic). This would suggest that the problem of inflation is a meta-problem – located in the meta-hierarchy. Perhaps economists can be persuaded to search the structures in N + 1 for the mechanism involved. The causes, of course, will be found only in H + 2!

And if a meta-hierarchy arises naturally in these contexts will there not be a *meta-Self*? Such a 'person' will be above all the Selves we have so far considered – the (N — 1)-Self, the N-Self, the (N + 1)-Self, etc. And this meta-Self will be able to alter the hierarchy of Selves, since it will contain covers of those selves. So somewhere in the meta-Self will be found the power which heals the body (for how can the tissues mend – becoming different tissues – unless the meta-Self controls the structure?), drives the mind on to master new areas of knowledge, and changes the personal and social relations between people.

Finding the meta-Self is usually the ambition lying behind all Eastern practices of meditation and other spiritual exercises (when it is sometimes called the Overself). Such achievement naturally leads to a certain detachment from worldly matters, from things and objects, from morals and ethics, from politics and economics – because all these things are part of the Selves (in the hierarchy H), whereas the meta-Self is in the hierarchy H + 1.

What would it be like to find one's way to the meta-meta-Self, I wonder?

5. Warping the Geometry

Now that we have seen how relations between sets of entities (provided those entities properly constitute 'hard' data) naturally give rise to a geometrical representation, in a multidimensional geometry, it is possible to study such relations directly in terms of that geometry. So it becomes natural to think of some social structure (such as is created by relations between people and institutions, for example) directly in terms of some geometrical ideas in, say, the space E^n (where we understand that, when the question is put, we can find the value of n precisely and also represent the geometry with all its subtle connectivities on a suitable computerized matrix). When this value of n is large (or even only larger than 3) we must turn to the computer to find our way around the specific geometry we have discovered – because we find it difficult to do this in our heads, however clever we happen to be. But then again we need not fear that in some way we are surrendering our human superiority to some damned machine – for we retain that meta-position already mentioned. Whereas the computer may be the one who finds his way around the geometry, *we* are the ones who know *why* the computer does just that – and of course that is why *we* must retain the ability to programme the machine to study the multidimensional structure.

If these geometrical structures we are considering arise naturally via the observation of hard data, then presumably we shall find (and have so found) them in the physical sciences. If it is the fact that *all* our rational human knowledge is manifest by our recognition of these *structures*, then that recognition will have been built into all our present-day scientific 'pictures' (although much of this will be expressed in mathematical symbolism); now, in order to see where to proceed with our new geometries, it will be helpful to look at some such (hard) scientific analogue of this methodology. We shall find thereby that *changes* in the geometry, peculiar to some specific study, are natural descriptions of what we usually call 'forces'.

Back in 1870 the English mathematician W. K. Clifford made the nice point that a worm living in a circular tube (like the inside of a metal ring doughnut) would experience a permanent sense of a force pressing him around in the circle. The worm's world would apparently include an unavoidable centralizing pressure which the worm would probably attribute to some external agency, since he would be unaware of the peculiar geometrical structure which he inhabited. This idea that a force can be 'blamed' onto the properties of the underlying geometry of the 'world' was later on (1915) incorporated into Einstein's general theory of relativity. Clifford's worm experienced a sort of *gravitational force* just as we do on this planet. We call it a 'force' and look for some external cause of it because we have a simple-minded view of the geometry of the world we inhabit (that view is the inheritance of ancient Greek geometry). Einstein's theory was built on the idea of changing that geometry (that E^3) so that it became 4-dimensional. The extra dimension which he introduced (mathematically, of course) gave him elbow room for bending the old 3-dimensional space into a big curved shape so that, just like Clifford's worm in his curved tube, we can then understand the experience of the force of gravity as something built into the structure of a new 4-dimensional space (an E^4).

If we cut through Clifford's doughnut (being careful to miss the worm) and open it out into a long straight circular cylinder, then the naïve worm will think that the force of gravity has entirely disappeared. If we then play a mean trick on him and keep bending the tube in different directions at different places along its length the poor worm will be overwhelmed with a confusion of forces coming at him from all directions. When he is moving serenely along inside the straight cylinder, and we have stopped twisting it about, the worm will think, 'How simple is the space I live in – it is 1-dimensional, just like Euclid said it was – here there are no obstacles to my moving along it and no awkward forces to throw my muscles into confusion.' But when it is bent back into a ring-shaped doughnut the worm would ask, 'Now some external force has been introduced into my world – I know the space is just the same as it always was, having learned all about it at school, so what could have caused this to happen?'.

Probably some extra-bright worm, called New Ton, would come up with the idea that the force is really caused by some new body appearing in the distance and that if the worms would just assume a certain mathematical law of attraction between bodies, then he could explain all the phenomena – including, of course, the exact shape of the circle which all worms must follow. The philosophically minded worms, whilst admiring New Ton's ingenuity, might still feel concerned by the implications. For this sort of explanation needs new ad hoc hypotheses for every new experience of force in the worms' world. The notion that the geometrical *structure* contains the source of these forces is, in contrast, a unifying idea. But then it also requires another dimension, so it places an intellectual burden on the worms. No worm can understand the Clifford–Einstein idea without extending his consciousness to the outside of the doughnut, by lifting his mental horizons from E^1 to (at least) E^2. But the idea also involves the notion of 'many possible geometries', and this is really a *meta-idea* for the worms.

Our modern human condition seems to suggest a similar dilemma. Einstein's theory of relativity, in its mathematical details, only tells us about physical bodies which are actually tiny pinpoints. Its gravitational theory is all about tiny points called Sun, Earth, Mars, etc., moving around in a modest 4-dimensional space. It is not about the sort of spaces we must inhabit when we argue politics, wage wars, make love, educate our children, or paint our pictures. But we have argued that these latter experiences are structural, that they need precise multidimensional geometries for their realization, and in that sense their structures are as significant as are the Clifford–Einstein ones. The idea that *forces are the result of warping the geometry* is just as profound and relevant in our social-personal structures as it is in the physicist's study of the gravitational field. Indeed it is even more emancipatory, liberating us from being confined by a low-hierarchical level of perception – without the elbow room provided by a meta-hierarchy. It 'lifts us up', it 'widens our horizons', it 'adds new dimensions' (literally) to our lives. Without it we cannot see ourselves as others see us, we cannot rise above our petty problems, we cannot laugh. Without it we can only blame 'others' (or 'them') for our difficulties – even when that is not rational – there will always have to be a New Ton to encourage us to look outside for that sinister agent who is 'causing' the unpleasant stress, whom we can piously blame for our troubles. For in the social-personal world *we are the structure* (the 'we' being our various collective Selves). If the structure changes, we change; warping the geometry means warping you and me, means warping our social institutions, warping our friendships, our working environments, and our creative abilities.

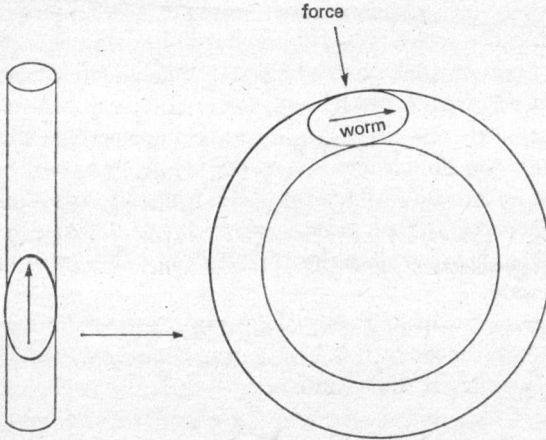

Fig. 14   Warping the geometry for Clifford's worm

But what is it that most directly experiences these structurally induced forces? Well, in the first place it will be the 'things' which are represented by the structure. It will be our friendship which will experience stress if the geometry of our relation changes. It will be the community which will be under tension if the geometry of its defining relations is being warped. But these basic structures, the geometrical backcloths (at various hierarchical and meta-hierarchical levels), also carry lots of appropriate traffic. This traffic must also experience the induced forces as the backcloth changes, like Clifford's worm (which is traffic on the backcloth structure) as the right circular cylinder changes into a torus (doughnut). If the relation between our two personal structures provides the backcloth for our friendship, that backcloth (which is in a suitable multidimensional space) carries the traffic which is a manifestation of our friendship – traffic which constitutes the communication between us, such as the mutual transfer of ideas, plans and action. As the backcloth changes by becoming smaller dimensionally (let us suppose), then its geometry loses vertices and the consequent connectivity properties. The first inkling we have of this process is the failure of certain kinds of traffic. Our structures 'back off' in some parts of the space E^n and an obstruction to the flow of traffic is felt in that region. So our friendship cools somewhat; perhaps you find someone else who is warmer, more attractive, more in harmony with yourself – more highly connected to your own structure (for that is what these various phrases mean). What is left of our friendship may well survive but it will be at a lower-dimensional level; we shall be connected by fewer vertices in our two geometries. The rapport between us, the dimensionality of the connection, will be reduced; we might be left only with what passes for politeness – the 2- or 3-dimensional traffic shared with almost everyone, the connectivities which allow exchanges about health, weather, and platitudes about society's welfare.

The forces which are associated with changes in the structure, and which are directly experienced by traffic on that structure, will also be peculiarly allied to different dimensionalities. For if the backcloth structure at some N-level in the hierarchy (call the structure S(N)) contains a number of tetrahedra (3-simplexes) and if two or three of these collapse into triangles (2-simplexes) by reason of vertices being destroyed, then any traffic which existed on those particular tetrahedra, prior to the warping of the geometry, must suffer a dimensional shrinking. The 3-traffic, which can only live in pieces of 3-dimensional geometry (tetrahedra), must now do one of two things when faced with this structural change. Either it must immediately move through the structure to some new haven of 3-dimensionality, or it must change its nature and become genuine 2-traffic. The first is rather like the toothpaste which moves out of the tube, when that is squeezed flat, into a new piece of 3-space. The second is much more common in our own social context where we 'lower our sights' and settle for some lower-dimensional mode of living. As individuals we can also be toothpaste, of course, when, for instance, we move from one environment (which has become dimensionally too small) to a new one – like when a man changes his job for 'more responsibility', or when a woman refuses the continued dimensional restrictions of being a 'housewife' and expands out into the bigger social structure for which (as traffic) she feels a deep need, or when a prisoner escapes from gaol. Is it so surprising that there should be so many common analogies in our language? 'I feel trapped inside this prison (home, office, marriage ...)', or 'I must get away from this restrictive environment', or 'I want to expand, to see the world', or 'I have no prospects where I am, I must leave' – all simple expressions of what we might call the *toothpaste syndrome*.

In comparison with this 'lowering our sights' in order to continue to fit into the new warped geometry is the classical problem of *compromising*. But often this is inevitable because the geometry offers us no way out, nowhere to go (someone has squeezed the tube without taking the cap off). But the *feeling* of 'having to compromise' is a painful one. It is the feeling of stress induced by the warping of the geometry, the direct experience of structurally induced forces. If it is a case of 3-traffic having to change its dimensionality because some 3-dimensional pieces of the geometry have vanished, then we speak of the traffic as experiencing (the geometry inducing) a 3-force. In a general context, if t-dimensional pieces of geometry are affected in the structural changes, we can speak of a *t-force*. Notice that we are giving the letter q a rest; but t stands for any of the numbers 0,1,2, . . .

Another example of a 3-force is our own gravitational field, if we adopt the Einstein explanation of it. That 3-force is experienced by all physics-type bodies in a 4-dimensional structure which was mathematically constructed by Einstein. You and I experience that 3-force at all times (except when we go out of the geometry and become astronauts) because we are usually examples of appropriate (inanimate) 3-traffic. But there are apparently well-documented accounts of particular persons who have achieved a non-gravity condition (disconnected themselves from the structure) and startled everyone near them by levitating (for instance St Francis of Assisi, St Catherine of Siena, and various other Christian saints – not least of whom were St Alphonsus Lignori, who rose several feet into the air while preaching to a large congregation, and probably the most famous of all, St Joseph Cupertino, whose levitations were apparently spectacular). This idea of disconnecting the geometry to produce anti-gravity was also the theme of *The Time Machine*, by H. G. Wells. Of course, serious scientists don't go along with these tales of levitation (as physical events), but if they were true we would need to understand how the gravitational warping could be locally unwarped by some animate being. If we cannot see the existence of real multidimensional structures then we shall certainly never understand how higher-dimensional simplexes can exist, or be made to exist, and which can somehow fill in vanishing 3-simplexes – to counteract the effects of lower-dimensional t-forces.

But physics-type gravity is a mundane illustration of a 3-force. Our personal, social and economic structures afford many more examples of directly experienced t-forces, due to changes occurring in the backcloth structure.

An example of t-forces on employment traffic

During the three years between 1974 and 1977 I was directing a research project, under the financial auspices of the Social Science Research Council, into the applications of this multidimensional analysis into the resources and structure to be found throughout the whole region of East Anglia. This region comprises the old counties of Cambridgeshire, Norfolk, East Suffolk and West Suffolk. After the local government reorganization of 1974 the two Suffolk counties became one, Suffolk, and Huntingdonshire was annexed to Cambridgeshire. But the study provides interesting and concrete examples of geometry-warping and structural t-forces. The hierarchy, H, for the geographical data sets seemed to need four (and potentially five) levels and these are indicated in the following list.

H:	Level	Set vertices	Number of vertices
	N + 3	East Anglia	1
	N + 2	Cambridgeshire, East Suffolk, West Suffolk Norfolk	4
	N + 1	Local Authorities (LA)	72
	N	Civil Parishes	1184
	N − 1	Wards	not counted

I shall illustrate by way of one county, Cambridgeshire, and in order to do this at the (N + 1)-level we need the following list of that county's Local Authorities (nowadays they are called District Councils). In that list below we remark that MB denotes 'municipal borough', UD denotes 'urban district', and RD denotes 'rural district'.

Cambridgeshire: (N + 1)-level vertices.

1. Cambridge MB	7. Chesterton RD
2. Chatteris UD	8. Ely RD
3. Ely UD	9. Newmarket RD
4. March UD	10. North Witchford RD
5. Whittlesey UD	11. South Cambs. RD
6. Wisbech MB	12. Wisbech RD

These areas, which naturally act as a cover and a partition of the county, are shown in Figure 15.

In addition to this rather obvious geographical data it was important to take into account government information (obtained from the Central Office of Statistics and Censuses) about the distribution of industrial employment categories (using what is called the Standard Industrial Classification of jobs, the SIC), and about Employment Status classes (the ES). Both of these new sets of things exist at different hierarchical levels, but here it is sufficient to identify the sets at the (N + 1)-level (where we can relate them to the (N + 1)-level pieces of geography). The two sets, (the SIC and the ES) are listed below.

(N + 1)-level Standard Industrial Classification (SIC) of employment:

1. Agriculture, Forestry, Fishing	15. Clothing and Footwear
2. Mining and Quarrying	16. Bricks, Pottery, Glass, Cement
3. Food, Drink, Tobacco	17. Timber, Furniture
4. Coal and Petroleum products	18. Paper, Printing, Publishing
5. Chemicals and Allied industries	19. Other manufacturing industries
6. Metal manufacture	20. Construction
7. Mechanical Engineering	21. Gas, Electricity, Water
8. Instrument Engineering	22. Transport and Communications
9. Electrical Engineering	23. Distributive Trades
10. Shipbuilding, Marine Engineering	24. Insurance, Banking, Finance
11. Vehicles	25. Professional Scientific Services
12. Metal Goods	26. Miscellaneous Services
13. Textiles	27. Public Administration and
14. Leather, Leather Goods and Fur	Defence

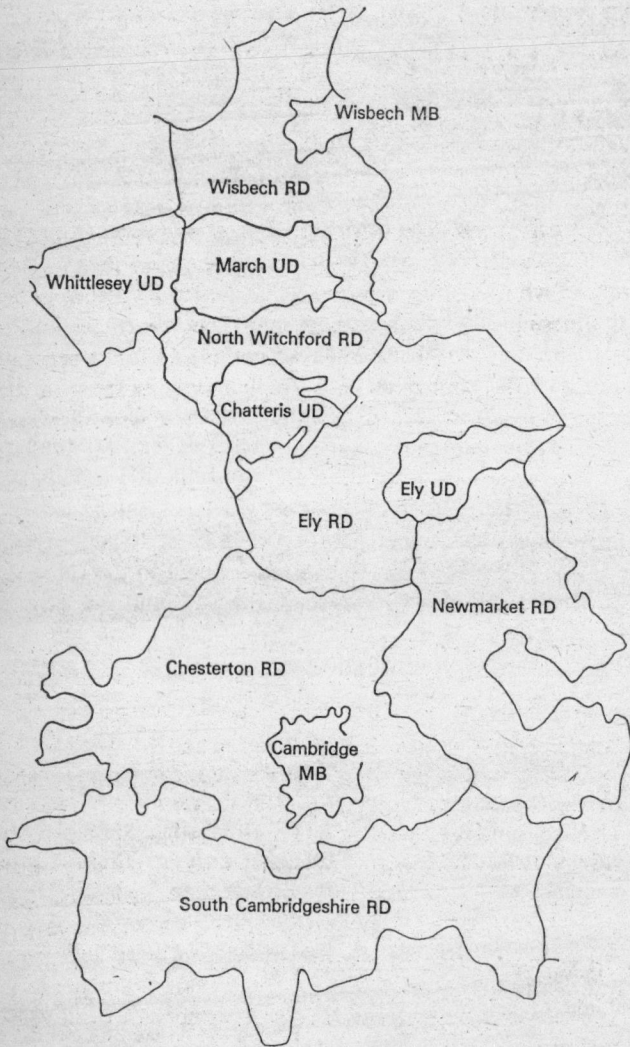

Fig. 15 *(N + 1)-cover of Cambridgeshire*

(N + 1)-level Employment Status classes (ES):

1. Self-employed (without employees)
2. Self-employed (with employees)
3. Managers (large establishments)
4. Managers (small establishments)
5. Foremen (manual)
6. Foremen (non-manual)
7. Apprentices, Articled Clerks, Formal Trainees
8. Professional employees
9. Family workers
10. Other employees (excluding professional)

The 1966 Census provides us with a table of numbers (a matrix array) whose rows refer to the twelve Local Authorities in Cambridgeshire and whose columns refer to the twenty-seven categories of the SIC. These numbers tell us how many people (male and/or female) were employed in the separate categories and in the separate authority areas of the county in 1966. Similar data exists from the 1971 Census; the same applies to data which refers to the Employment Status classes and the authority areas, both for 1966 and for 1971. In all cases it is possible to obtain separate data for males and for females.

Out of any one such table we can derive a large number of genuine relations (with matrix arrays consisting of 0's and 1's). One such way is to find the average along each row (a separate average number for each Local Authority in the county) and then to replace each entry in the row with a 0, if that entry were less than the row's average, and with a 1 otherwise. The resulting relation is between the set Y, of Local Authority areas, and the set X, of SIC job categories, at this (N + 1)-level.

Now it so happened that the resulting structure, KY(X), using the data for males, told us that, for example, the city of Cambridge (element (1) in the set Y) was a 6-simplex in 1966 and a 7-simplex in 1971. In the 1966 structure the town of highest dimension was Wisbech MB, being a 9-simplex, and in that year the city of Cambridge was a 6-dimensional face of Wisbech. This meant that 6-traffic would flow freely in the geometry between the two towns but that 7-traffic in Wisbech could not move to Cambridge. Some examples of what we could contemplate as traffic on these SIC-based structures, the KY(X), are the following:

1. Numbers of employment-seeking males/females: then e.g. 2-traffic means people who are eligible for three kinds of job.

2. Numbers of unemployed persons: then 0-traffic means people who can only do one job.

3. Private houses built in each area: then 4-traffic means houses readily accessible to five kinds of work.

4. Commercial buildings: then 5-traffic means buildings which locate six job categories.

5. Volume of vehicular traffic flow through each area: then 0-traffic means vehicles which are connected with only one kind of work.

6. Goods imported/exported by each area: then 3-traffic means goods which are involved with four industrial activities.

7. Financial investment per area: then 1-traffic means capital available for investment in two types of employment.

8. Taxation levied per area: then 7-traffic means tax revenue derived from businesses which simultaneously appeal to eight categories of work.

9. Investment in education in each area: then 1-traffic means investment in education which is concerned with two types of industrial training.

10. Investment in health, social amenities, welfare per area: then 3-traffic means investment in health clinics which deal with medical problems peculiar to four kinds of workers.

The typical relation we have just described, using the row-average as a norm or base, will provide us with a backcloth structure $S(N + 1)$ which corresponds to a uniform distribution of employment activities throughout the county. This correlates highly with a routine Land-Use survey which identifies the locations of factories, schools, amenities, farms, forests, etc., in the county. Since in 1966 Wisbech MB was the highest-dimensional simplex in this backcloth (using the data for males, with respect to the SIC), we read it as meaning that Wisbech had the largest number of above-average employment opportunities for males. By going to the conjugate complex $KX(Y)$, in which the simplexes became the $(N + 1)$-level job types, we found that the dominant element was Construction (as an 11-simplex), closely followed by Agriculture, and Distribution (as 10-simplexes). These are the job types which were distributed most widely throughout the county, which gave the *feel* of the county. This also means that a prospective agricultural worker looking at Cambridgeshire in 1966 would constitute 10-traffic on the structure $KX(Y)$. He would have simultaneous opportunities of eleven areas of work in this above-average structure. If he were already working in one of the county areas then he was also traffic on KY(X) and his *mobility* in that geometry (from one area to another) would be dependent on the connectivities between his area and the others.

Now we can collect the results of the research analysis and show how the various structures changed between 1966 and 1971. These changes induced t-forces throughout the county, forces which must have been felt by many individuals working therein.

Figure 16 shows the way that the q-values (dimensions of the q-simplexes) of the Local Authority Areas in Cambridgeshire changed, in KY(X), between these two census dates. The structure refers to the SIC data for males and the *change* in the structure is denoted by the symbol $\triangle S(N + 1)$.

We notice that Wisbech MB changed from being a 9-simplex in 1966 to being a 7-simplex in 1971. This means that 9-traffic or 8-traffic in Wisbech would experience t-forces of repulsion; such traffic would be squeezed out, if it had anywhere to go. We see that there was no 9- or 8-simplex in the county in 1971, so the traffic had nowhere to go – therefore it had to compromise and reduce itself to being 7-traffic. In fact, this high level of dimension means that it would not easily be noticed by employees, employers, or investors – for how much capital has been earmarked for simultaneous investment in ten kinds of industrial activity? Nevertheless the t-forces are there and, being repulsive, they can well be the beginning of more noticeable stresses (at lower dimensions) later on. It is also to be expected that t-forces which are difficult to detect are also likely to be completely overlooked by our professional public servants, making it more difficult for them to accept even the gist of these concepts.

In comparison with Wisbech MB we see that Wisbech RD and Ely RD both experienced t-forces of attraction, as their q-values increased. In each case we can see this as a measure of the widening employment horizons in those rural districts.

The next two diagrams, Figures 17 and 18, show comparisons between the structures vis-à-vis male and female data; the first refers to 1966 and the second to 1971 (both using the SIC categories).

The warping of the geometry we are now considering is the change from a structure KY(X; Male) to a structure KY(X; Female).

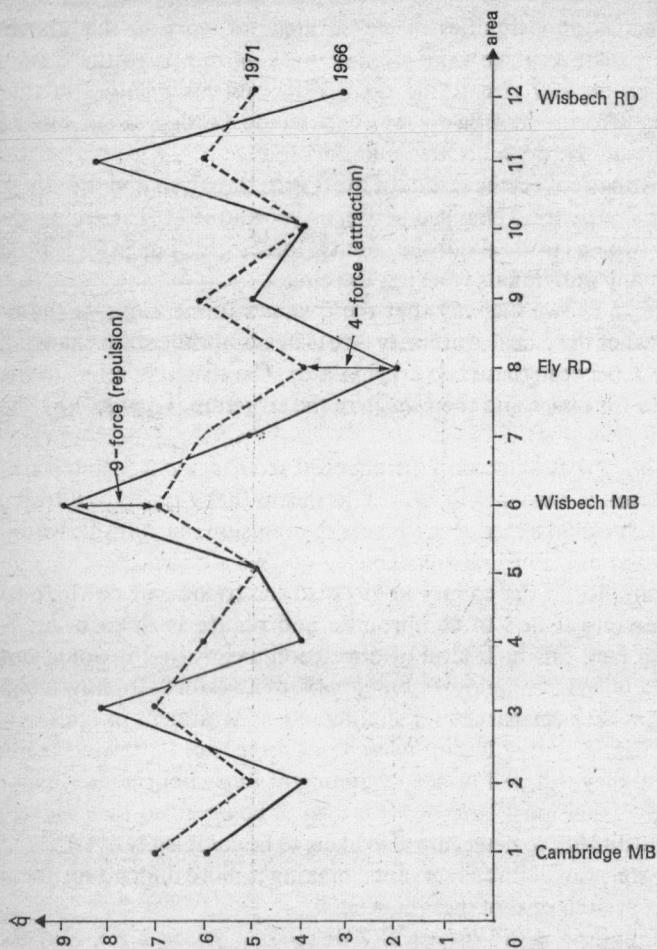

Fig. 16   Cambridgeshire – t-forces due to $\triangle S(N + I)$

The vertices, X, are the same and the Y's refer to the same set of areas, but the structures are appropriate to Male-oriented traffic or to Female-oriented traffic. We notice in Figure 17 that the structure corresponds to greater above-average work opportunities for Men than for Women. It is particularly striking in the two main towns, Cambridge MB and Wisbech MB. Whereas Female-oriented traffic, of 4 dimen-

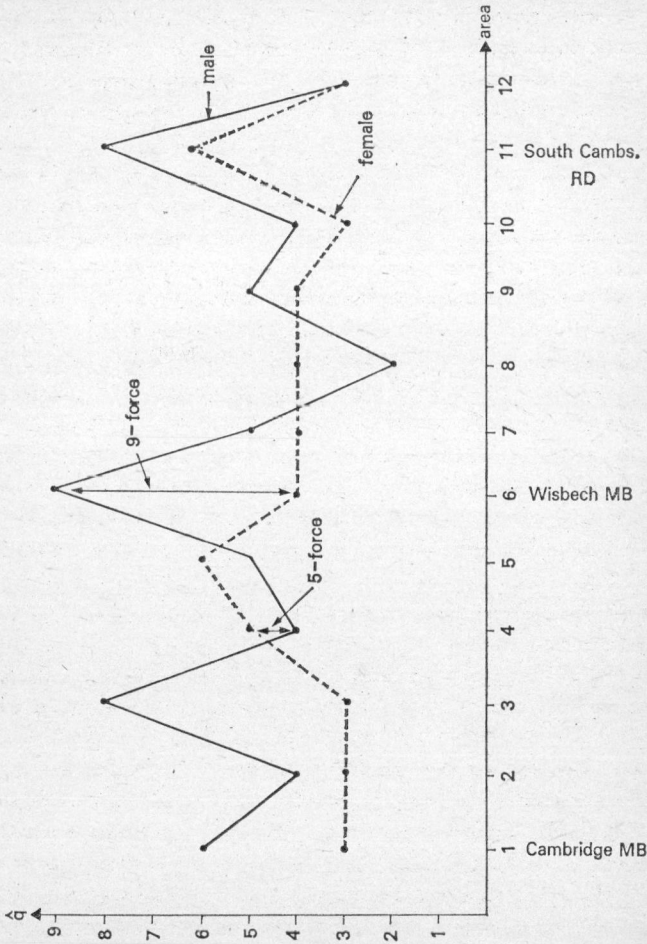

Fig. 17 Female–Male t-forces 1966 Data

sions, could only exist in Wisbech MB, Male-oriented 9-traffic would feel at home there. If you were a Male person in Wisbech on one Sunday in 1966 and you woke up on the Monday only to find that you were suddenly Female, then you would experience t-forces of repulsion in that warped geometry – for all values of t from 5 to 9. Your horizons would shrink about you as the geometrical walls closed in.

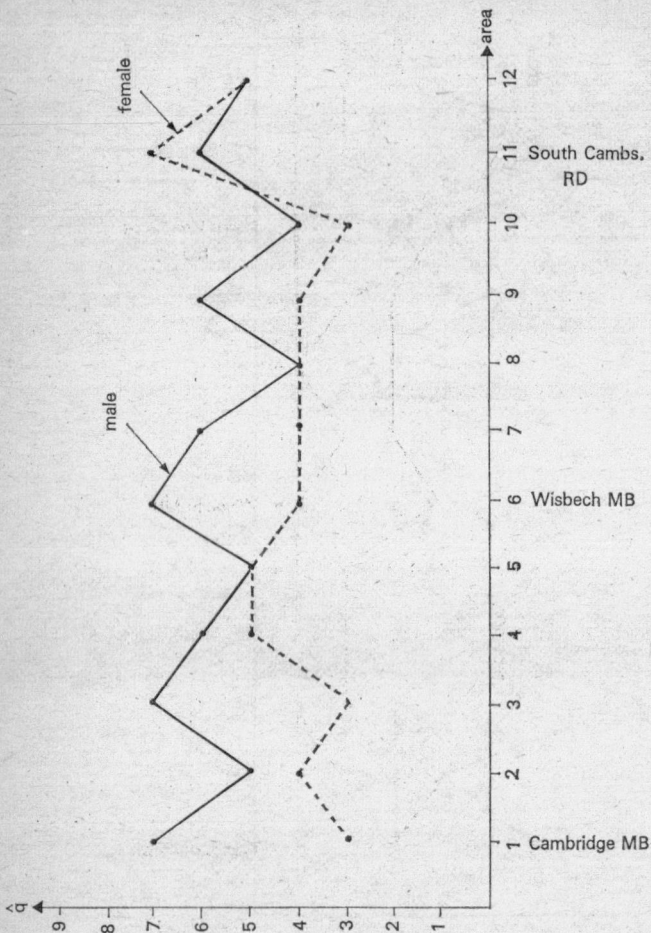

Fig. 18　Female–Male t-forces 1971 Data

You would most probably lose your job, money would not be easy to borrow – if at all – and many people you knew would not want to help. Only in Ely R D would you feel the expansive sense of new opportunities, of t-forces of attraction, and to a lesser extent in March and Newmarket.

In Figure 18 we see the similar warping of the geometry for the 1971

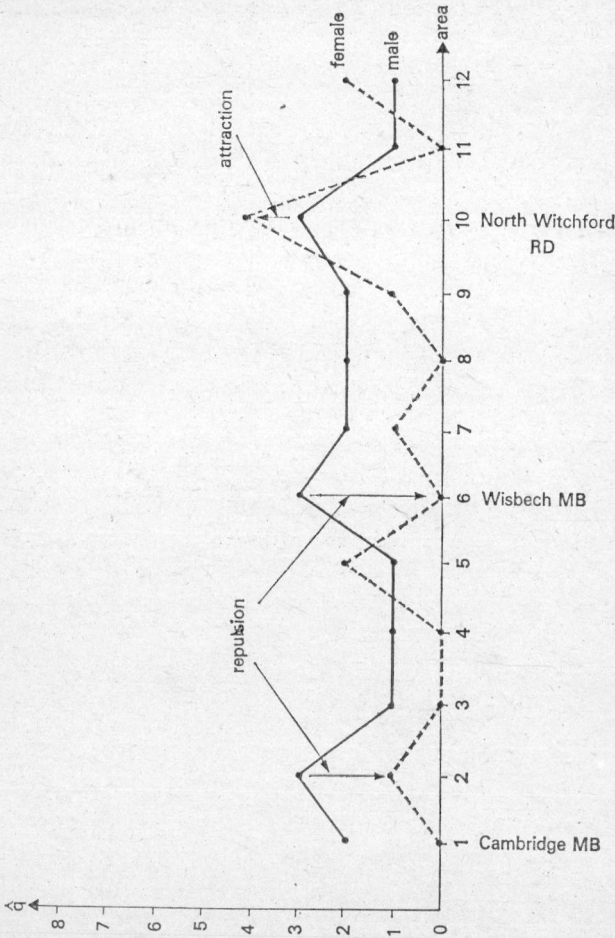

Fig. 19   Female–Male t-forces ES Data 1966

structures KY(X; Male) and KY(X; Female). But now it was getting worse! Only South Cambridgeshire RD offers any improvement, while throughout the rest of the county the Male-oriented structure is dominant.

But perhaps women fare better in some other structure, for example in that based on the relation between Local Authority Areas

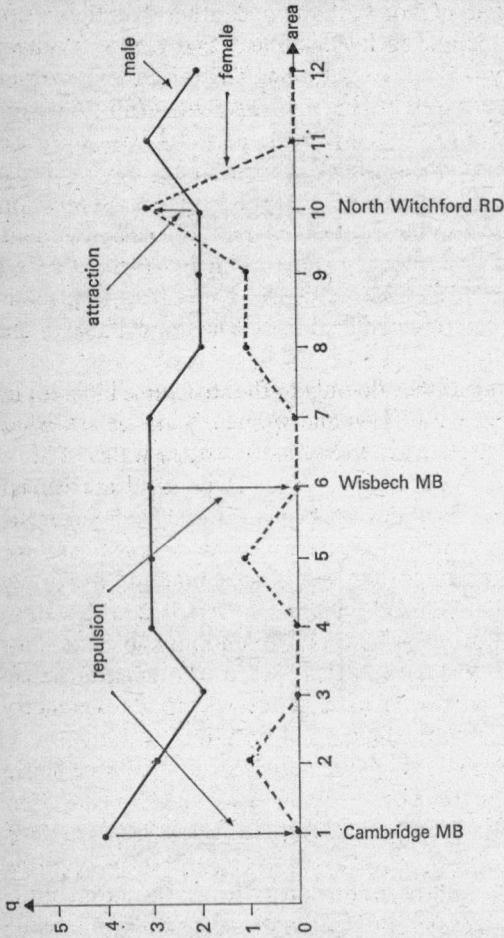

Fig. 20   Female–Male t-forces ES Data 1971

and the set of Employment Status classes (ES)? The analysis is shown for 1966 and for 1971 in Figures 19 and 20 above.

The change from KY(X; Male) to KY(X; Female) produced widespread t-forces of repulsion in 1966 and these became worse in 1971. Only North Witchford RD showed an attraction in 1971, and what is more significant than in the SIC data, is the low dimension of the

Female simplexes. Values of 0 or 1 are the rule rather than the exception for these dimensions, and such values mean that KY(X; Female) can usually only carry 0-traffic or 1-traffic. The 0-level means that Females have only one choice of Employment Status (no prizes for guessing where that is in the status-stakes) and the 1-level means a choice of two. What price then being a woman in our employment-based structures? Why do we suppress them so – for it is suppression in the most literal sense – in terms of calculable t-forces consequent upon warping the multidimensional geometry which they inhabit (but which men do not inhabit)? It is as if the men are living in large spacious mansions while the women are confined under the low ceilings in the basement.

These results, applying as they do only to the structures inherent in the commercial employment of men and women, make us ask what comparable structural stresses are to be found in other walks of life – such as arise by way of educational, cultural, political and social relations? The deep trouble with the problem is that even when the Central Government, in its meta-role, passes statutory laws to enforce 'equal opportunities' or to forbid 'sex discrimination' it is only playing with the *traffic* on the underlying structure. If that structure, that backcloth, has the problem built into it (and no one can see it), then the laws become near farcical and practically unenforceable. But of course the alternative is to effectively warp the geometry of society and that is usually known as revolution.

Another aspect, which might appear paradoxical, to all this lies in the ambivalence of the relation between men and women. For, certainly at a personal level, there can be great affection between them and tenderness and generosity, but in this work context the opposite seems to be the case. The difference obviously lies in the nature of the hierarchical levels of their relationships, as we have already discussed in Chapter 3. For the personal intimacy is appropriate between the man and woman when each is the N-Self (and that intimacy can then automatically involve the $(N-1)$-self, the $(N-2)$-Self, etc.). But the work context, as we have analysed it, refers to the $(N+1)$-Self, or the $(N+2)$-Self, or the $(N+3)$-Self, and so on. Then the suppression of woman is actually the suppression of the collective $(N+2)$-woman by the collective $(N+2)$-man, etc. This of course makes it harder for the oppressed to bear, because t-forces which are manifest by the warping of appropriate $(N + 2)$-geometry cannot avoid filtering downwards into the $(N + 1)$-geometry and the N-geometry, by reason of the relations which link the cover-sets together. So the stresses felt by any oppressed subset of society cannot be successfully fought by the individuals, as N-Selves. They can only be fought at the highest level, say the $(N + 2)$-level. This is why 'politically conscious' women launch campaigns to unite their sisters in some 'mass' movement. After all, the 'mass' movement is an attempt to create the collective $(N + 2)$-Self, and it must be true that therein will lie the possibility of successful 'political' (more $(N + 2)$ relevance) action. Nor will this sort of thing be peculiar to movements aimed at 'liberating' women in our society. In any set of relations where some subset of people find their geometries dimensionally inferior to those of others, there will be the same sort of situation: 'oppression' in some form or other (to a greater or lesser degree), expressible in terms of t-forces of repulsion between warped geometries. (See reference[18] for an account of social studies done in comparing the male–female struggle with that of the white–Negro struggle in America.) But this hierarchical point makes it easier to understand why there is often a deep inertia in society to introduce such changes as we imply. The inertia expresses the sense of bewilderment (often incomprehension) about 'the problem' – by people who can only see society through the N-Self (where often there genuinely is no problem).

Pain and pleasure associated with t-forces

You will have noticed that we are using the word 'warping' in a wider sense than is to be found in common usage, for we are including in it the idea that the geometry can change in quite startling ways. These include both the loss of geometrical points (vertices), with its resulting loss of connectivities in the structure, and the accretion of new points, with the consequent extension of existing connectivities. Indeed any twisting and turning, or stretching, of a complex K – and this corresponds to an ordinary meaning for 'warping' – which is not accompanied by any changes in the connectivities therein, does not introduce any t-forces into the structure (chiefly because our geometry is not dependent on ordinary ideas of distance – a mathematician would say that the topology is non-metric). It is therefore natural to ask how such changes can come about, or where we can find the seat of that action which initiates them.

Since the warping of the geometry means that the complex K changes into a distinct complex K' it means that there is a change in the *relations* which define the structure (and this can mean changes in the notional sets Y and X as well). But since the system we might contemplate will be described by a hierarchy, H, of such sets and relations, it follows that changes amount to changes of H. The seat of any action which can initiate such changes must therefore lie in the meta-hierarchy, H + 1, as we have previously described. This need not be the end of the story, of course, because if we need to contemplate action which can change that action which changes H, then we would need to look at the meta-meta-hierarchy, H + 2, and so on. But does that mean the process is potentially never-ending? Yes, it does so mean. This must be both the price we pay and the delights we enjoy for being animate and human. In contrast, we notice that the structure of the world which has been explored and exposed by the physical scientists is supposed to be unchanging – it knows no meta-world where the rules are being laid down. Perhaps this is what makes scientific theories both attractive (by suggesting a secure and permanent structure) and repulsive (by suggesting that development and change is impossible, that all is dead)? But until the idea of structural changes can be accepted as a legitimate factor in scientific endeavour it is surely unlikely that the concepts and theories of the physical sciences will be adequate as a basis for the pursuit of social science? In practice this means that the conventional idea of the 'law of nature' is too rigid – because it implies the non-existence of the meta-world, where 'laws' can be changed. It looks rather as if the so-called 'laws' of social science will have to be sought for in the meta-hierarchy (when they will be statements about the rules which govern how rules can be changed) or even, of course, in the meta-meta-hierarchy (when they will be statements about the rules which govern how the rules which govern how the rules can be changed can be changed!).

But for our present purposes, with these heady thoughts floating in the wings, we can pursue the interpretation of warping the geometries of H without always having to explain how it has happened, or even why it should happen. In this context we can remind ourselves of the suggested hierarchy relevant to our collective Selves (see page 62). There we have already used an H which notionally went from $(N + 3)$ down to $(N - 3)$, with the commonsense idea of 'self' being referred to as the $(N - 1)$-Self. We will identify this hierarchy via the following scheme:

H.	Level	Descriptive features
	$(N + 3)$-Self	The nation, or similar ethnic group
	$(N + 2)$-Self	Subsets, e.g. political, economic, cultural, identification of
	$(N + 1)$-Self	Smaller subsets (towns, cities), work groups (unions), play groups, etc.
	N-Self	Family subset, friends groups, smaller work groups, play groups etc.
	$(N - 1)$-Self	Conscious individual self, personal interests, work, appearance, culture
	$(N - 2)$-Self	'Pieces' of the $(N - 1)$-Self, in tastes, behaviour, appearance, modes of speech
	$(N - 3)$-Self	Unconscious self; vertices cannot be known consciously – probably intuitively
	$(N - 4)$-Self	Deeper unconscious self, if possible

We do not need to specify the fact that these levels are all linked by suitable relations between the appropriate sets. These relations naturally define a great many complexes (which however may all be united as a collection which itself will be a complex – if it is helpful so to think of them) with their own peculiar multidimensional geometries. So, for example, there is a relation (and two sets, called Y and X, let us say) which brings the $(N - 1)$-Self and the $(N + 2)$-Self together: the set X can contain those elements which properly belong to the $(N - 1)$-level whilst Y is a list of elements describing the $(N + 2)$-level. This means that we have two complexes, KY(X) and KX(Y), the one being called the conjugate of the other. In the structure KY(X) my $(N - 1)$-Self is a vertex (since X lists all the conscious selves as individuals) and the $(N + 2)$-Selves are then simplexes built out of these vertices. So in KY(X) I am personally just a 'point' in the geometry–a nobody in the sense of having no structure, a mere pawn in the game. But at the same time, and with as much significance, in the other structure KX(Y) my $(N -)$-Self is a simplex (a polyhedron) whose vertices are the bigger words to be found in the set of $(N + 2)$-Selves. So then I am personally very important and my connections with other persons is what holds the $(N + 2)$-Selves together. But this is an ambivalence which we all feel, that we are 'one of the crowd' and at the same time that we are 'more important than being mere pawns'. People who over-use the statistical approach (and this includes all our present public servants and politicians, as well as many of our social scientists) are working exclusively within the structures like $KY(X)$ – where we individuals are merely the points of the geometry. But since our deeper instincts, as well as our conscious minds, see our structures in $KX(Y)$ there is a permanent conflict with the mandarins. Furthermore, warping the geometry which represents this relation inevitably introduces t-forces which are felt by the $(N - 1)$-Self as well as by the $(N + 2)$-Self. So we find ourselves saying things like, 'the workers are feeling the pinch' – which could mean for instance that some $(N + 1)$-Self (which would be 'the workers' in some context) is experiencing some t-forces of repulsion due to the changing relation of H. At the same time as this is happening, it might equally well be true to say that 'John Doe doesn't know that times are hard', if only because those t-forces which by implication have found their way into $KX(Y)$ have, it so happens, missed the simplex which is John Doe.

This means that warping the hierarchy H has repercussions, via various t-forces, all the way up and down. Sometimes those t-forces are forces of *repulsion* (because some parts of the geometry have collapsed – forcing out the toothpaste) whilst sometimes those t-forces will be forces of *attraction* (because some new accretions or extra connectivities have occurred, allowing traffic to move more freely than before, or allowing higher-dimensional traffic to exist where it previously could not).

Let us examine the possibility that t-forces of repulsion are recognized as *pain* (by 'the structure') whilst t-forces of attraction are recognized as *pleasure* (by 'the structure').

If human beings are 'pleasure-seeking' we would therefore describe them as searching after an expanding geometry, and they must have the capacity to find it (or to make it expand) by themselves moving into the meta-hierarchy where the action can be initiated. But this expansion will be manifest at some or all of the hierarchical Selves, and it will be initiated at some or all of those same levels. Pleasurable sensations can be experienced by the $(N - 2)$-Self: they can be physical or psychical; they can induce pleasure upwards through the $(N - 1)$-Self to the family circle to the social group etc. When 'the nation is happy' it is because, in this sense, the individual people are happy. But we also see both the truth and the falsity in the idea that the nation is just a collection of individuals – since the $(N + 3)$-Self is part of a vast structure of precise geometry whose vertices are the $(N - 1)$-Selves (at least). It is the connectivities in that geometry which are also an essential part of the common meaning of 'nation'.

This idea of something possessing a sort of 'wholeness' and thereby being superior to its 'atoms' is of course the ancient idea that 'the whole is greater than the sum of its parts' – and was certainly an important part of Aristotelian philosophy. It has more recently had a resurrection in the theory of Gestalten, and Gestalt psychology has been an important part of what is generally called 'structuralism'.[19] The thesis of this book is clearly in harmony with this modern trend – but I believe it goes much further by insisting on the precise mathematical nature of structure under a general methodology, making it 'hard'.

Since the sort of structure we are talking about is of the same *nature*, whether it be at $(N - 2)$ or at $(N + 3)$, it immediately becomes feasible for any individual to be sensitive to each level – in theory, certainly. But this process might not be easy without psychic maturity or by some encouraging education. It will follow that particular persons might well find a role in H, whereby they *personify* the higher-level Self. So it is not surprising that The Queen, or The President, becomes the $(N + 3)$-Self for some set of people. In that person the individual $(N - 1)$-Selves can find a vicarious existence as an $(N + 3)$-Self. Similarly we find 'the leader' in many situations, fulfilling the same sort of role – as Louis XIV once said, 'L'État, c'est moi.' So we can probably put The Prime Minister as the personification of an $(N + 2)$-Self, the film star as an $(N + 2)$-Self, the Union Secretary as an $(N + 1)$-Self, perhaps the Shop Steward as an N-Self? But then the great personal danger [sic] to this individual who is carrying the whole load of being an $(N + ?)$-Self is that which can arise if he/she becomes detached from the other $(N - 1)$-Self, etc. For then the 'public person' is a trap which breaks up the geometry which used to connect him with his true $(N - 1)$- or $(N - 2)$-Self. On the political scene such identification has led many times to the arrogance and tyranny of the dictatorship – as Lord Acton's famous dictum runs, 'power corrupts, and absolute power corrupts absolutely'. And in other spheres there will be similar problems as people find their geometries shattered at the lower hierarchical levels, because they rise to become 'stars' in some context – witness the tragedy of the actress film stars who become $(N + 2)$-Woman and whose $(N - 1)$-Selves fall apart.

All this is the other side to 'pleasure seeking'; it illustrates the pain of disrupted geometry. And is it not the same sort of story as the Jungian idea of 'projection'? But here we can be discriminating, and see the idea as projection allied with specific hierarchical levels. Furthermore, we would expect that what we might project is the geometry in us which is not really conscious. So we would want to interpret Jung's *anima* or *animus* as representing geometrical structure which is in the $(N - 3)$-Self, or perhaps the $(N - 4)$-Self? This would certainly make it possible to understand the role of the *intuition* in this process – a vague awareness of something deep inside one which yet makes contact with what is conscious and outside. And it would be consistent to believe that the $(N - 3)$-Self has a geometry in some suitable E^n (where n is going to be large), in the same way as we can speak of and identify the conscious geometries of the higher levels. Of course if 'science' is to come up with the physics and chemistry of this $(N - 3)$-Self it will have to destroy it in the process – because it cannot be projected on to the $(N - 1)$- or $(N - 2)$-World without losing its own intrinsic geometry (since vertices at $(N - 3)$ will be covered by sets of the 'scientific' vertices consciously observed at $(N - 1)$).

But the warping of the geometry at the level of the Unconscious Self will produce the same sense of pain or pleasure at the higher levels as will a similar event being initiated at, say, $(N + 1)$. So is this the field where psychiatric theories find their ground, where Freud was saying that sexual disturbances (in the $(N - 1)$- and $(N - 2)$-Selves) produced permanent and deeper geometrical distortions in the $(N - 3)$-Self, and that these in turn would produce higher-level symptoms of neurosis? Did it make a big appeal to people because it awakened the intuitive sense of the $(N - 3)$-Self, rather than because it was somehow obviously seen to be irrefutable science? Because it seems very sensible to believe that the sex drives must be extremely strong in the race (which will be something like an $(N + 5)$-Self), then the only weakness in the Freudian approach (which, incidentally, was taken much further by Wilhelm Reich in his study of the orgasm) might arise from seeing *only* that cause in structural distortions. So we find reactions against that theory by Adler and Jung – not to mention the sort of panic which 'ordinary' $(N - 1)$-people can feel when they are made aware of powerful drives in their $(N - 3)$-Selves (when they had lived comfortable structural lives without having to admit the existence of such levels).

Now there is a structure appropriate to the $(N - 2)$-Self, where we can regard biological science as entrenched, which is the projection of the $(N - 3)$-Self (or of some of it). That projection is what we have already referred to as the mechanistic view of the human body and psyche; it is the natural result of trying to provide an explanation of the $(N - 3)$-Self in terms of physics and chemistry and their 'laws'. Of course it is 'true' in a very real but limited sense – but that 'true' expresses only a tautology which is inherent in the mechanistic methodology. It is a 'truth' which the conscious mind (the rational mind *qua* ego) must find acceptable – because it is a projection into the $(N - 2)$-World – but it cannot be described as either 'true' or 'false' in one's direct experience of the $(N - 3)$-Self. But it is likely to give rise to an *intuitive feeling* of being largely irrelevant. This intuitive feeling is the awareness of the $(N - 3)$-Self, of its geometrical structure in the hierarchical scheme of things. The physical pleasure of well-being (of being in 'good health') is an awareness of the expansive and adequate geometry in the $(N - 3)$-Self, whilst the physical pain which is manifest in the $(N - 1)$-Self can also arise from the disruption of the $(N - 3)$-geometry. These things raise important questions of 'cause' and 'effect' in medical science, not least the struggle between those who see only the soma and those who see always the psychosomatic.

When David Mulhall's patient, John, came to the clinic he was in pain (psychological-personal-social pain) and his wife was in pain, because the geometry of their union was inadequate – at various levels. The treatment consisted of making the patient conscious of the structure (which in effect was a lack of structure) of their united N-Selves. In that sense it was not a psychiatric treatment (which would have been more concerned with concepts which hopefully relate to the $(N - 3)$-

Self) but a psychological one which could be expressed in $(N - 1)$-Self terms *about the* N-*Self geometry*. As this began to have some effect, due no doubt to the sensitivity and intelligence of the patient, we have seen that the geometry changed into a condition where the connectivities (which encouraged genuine union between man and wife) became stronger, where the dimensions increased and allowed higher-dimensional traffic to flow (this traffic contained consideration, care and tenderness – all being relatively high-dimensional) between them. So the pain began slowly to give way to pleasure – man and wife found themselves happier living together. The process clearly involved the need for John himself to move into a meta-hierarchy position, so that he could share the diagnosis and the treatment with the psychologist.

The effect of drugs to produce analgesia might well provide an opportunity for the Self to come to terms with a new structure, at $(N - 3)$ or $(N - 2)$. But drugs could also be inhibiting the process, particularly if the pain is definitely due to the warping of the $(N - 3)$-Self. Then drug dependency would be inevitable. The role of addictable drugs must be one which seriously affects the $(N - 3)$-Self and its geometry: perhaps initially expanding some part of the geometrical structure which is ultimately at the expense of the rest? If that were the case then withdrawal symptoms would naturally be expressions of the pain of a shrinking geometry, when that new structure could not be maintained. See how this experience was described by Thomas de Quincey in his book *Confessions of an English Opium Eater*, as he conducted experiments on himself in an attempt to give up the drug.

... the symptoms which attended my case for the first six weeks of the experiment were these: enormous irritability and excitement of the whole system; the stomach, in particular, restored to a full feeling of vitality and sensibility, but often in great pain; unceasing restlessness night and day; sleep – I scarcely knew what it was – three hours out of the twenty-four was the utmost I had, and that so agitated and shallow that I heard every sound that was near me; lower jaw constantly swelling, mouth ulcerated ...

The $(N - 2)$-Self feeling the pain of the $(N - 3)$-geometry contracting?

Perhaps *sleep* is the condition wherein we directly experience our Unconscious Selves – $(N - 3)$, $(N - 4)$ ...? If those Selves are in pain due to warped geometry then it can be manifest to the Conscious Self only in $(N - 1)$-Self terms – such stuff as dreams are made of, or (with de Quincey) the inability to sleep at all. This might well be the total rejection of the psychic energy from the structure of the Unconscious Self (at the $(N - 1)$-level) into that of the Conscious Self. For we must bear in mind that it would be wrong to imagine that the Unconscious Self is *only* manifest at the lower-hierarchical levels, such as $(N - 3)$, $(N - 4)$. For one reason, the cover-sets which define the hierarchy ensure that $(N - 3)$-structure is contained in, for example, $(N + 1)$-structure. Then again, the existence of the conjugate structures means that vertices, at $(N - 3)$, are also manifest as simplexes, at $(N + 2)$ etc., by reason of the defining relations of H. So there is a sense in which changing to the conjugate effectively turns the hierarchy on its head. This must also mean that any Unconscious structure in the Self is not only manifest at each level in H but will also be part of the meta-hierarchy condition, $H + 1$. So we can expect intuition about one's structure (say, the sense of illness at $(N - 2)$) and also intuition about how it should be changed (the meta-sense about getting well again).

Anti-vertices and anti-Self

If the meta-Self is to function smoothly and efficiently it must be able to contemplate a selection of Selves (all up and down the hierarchy H). And in particular it would need, at the least, a structure for Self which is waiting in a sort of opposition condition – as a possible replacement for the functioning Self. This idea is really Jung's idea of the Shadow of the Self. But in this context we can see it as having a geometrical structure, say $KY(\bar{X})$, whose vertices are in opposition to the members of X. Such vertices will be called anti-vertices: if $\bar{X}_1$ is an anti-vertex for X_1 they must be mutually exclusive in a destructive sort of way. The meta-Self would then be able to choose whether Self should contain X_1 or $\bar{X}_1$, as two possible extreme opposites. Such a choice is actually a choice between different hierarchies, say H and $\bar{H}$, for Self. It is obviously a very limited choice, if that is the only choice, but because of that very fact we must expect not less from the meta-Self. Furthermore we would naturally consider an anti-set $\bar{Y}$ (for Y) when we are dealing with the conjugate structures.

It is then natural to regard the structure $KY(\bar{X})$, which we can derive from $KY(X)$ by assuming the same relation λ between Y and $\bar{X}$ as we previously had between Y and X, as the anti-geometry – corresponding to the initial geometry. Surely we cannot resist saying that this will represent the other, the Shadow, the anti-Self? Naturally the meta-Self will be able, when things are going smoothly, to swap around the vertex set between an initial set X and an initial anti-set X̄ in order to find some new acceptable set, say X′, which might contain some of each. That will then produce a current Self – which the meta-Self presumably finds acceptable (for the moment). If the set X is said to be a set of positive features then we would need to say that X̄ is a set of negative features; if X are attractive then X̄ are repulsive. And of course these differences can exist at each of the levels of the hierarchy H. So there can be an $(N - 1)$-anti-Self (the extreme opposite of my usual $(N - 1)$-Self); there can be an N-anti-Self (the opposite family-man, etc.); there can be an $(N + 1)$-anti-Self (the opposite or anti-factory, the anti-union, the anti-town, etc.); there can be an $(N + 2)$-anti-Self (the anti-political party, the anti-economic group, the anti-workers), and so on. And since X and X̄ are mutually antagonistic the two geometries KY(X) and KY(X̄) are also mutually incompatible – they cannot be joined or united by any natural process – *love* between Self and anti-Self is impossible (as we have described it so far). Indeed the only natural traffic between KY(X) and KY(X̄) is the opposite of identification: it must be the *anti-traffic* to love – otherwise known as *hate*.

But if the meta-Self keeps the anti-Self well hidden (does not call on it for any reason), then this hate will also find no expression – it will not in fact be needed. On the other hand, if the anti-Self is called forward into the Conscious Self because of some need which the meta-Self sees, then the hate will be manifest in all its familiar terms, expressing rejection, violence and destruction. This will be manifest again at some or all of the hierarchical levels of Self. So we can find the personal hate of the $(N - 1)$-Self, the group hate of the $(N + 1)$-Self, the nation hate (war) of the $(N + 3)$-Self. Then the struggle goes on until the meta-Self calls it off – but is Self ever quite the same again? Is it ever possible to hate another without hating one's Self – for do we not identify that other with one's own anti-Self?

Structure can be destroyed by bringing up the anti-structure so as to make suitable 'anti-contact' at certain vertices. Then the result is a new structure of lower dimension and inferior connectivities.

Victory in war comes about when the enemy's geometry has been sufficiently destroyed for him (his $(N + 3)$-Self) to make war – that is to say, go through the motions of making war – those motions being high-dimensional traffic on his own $(N + 3)$-Self. That happens when his towns are shattered, his armies are scattered, and his morale is in pieces: then his meta-Self has no choice but to call it off, and his $(N + 3)$-Self surrenders to his $(N + 3)$-anti-Self (which is the $(N + 3)$-Us!).

When woman is reduced to a low-dimensional structure, as sex object, it is achieved by the hate which destroys her Self – at all levels: and it is truly said that pornography is the expression of this hate.

When a child suffers an abominable education in such a way that his geometrical structure is reduced in dimensions (and probably his meta-Self is inhibited in its development) then he is the victim of hate – or the triumph of his mentor's anti-Self. Which is why wicked teachers never get better: they are fighting a continual war with an anti-Self, and seeing it work out on their pupils. Which is why that great imaginative educator, A. S. Neill of Summerhill School, wrote his books entitled *The Problem Parent* and *The Problem Teacher*: he never wrote one called 'The Problem Child'.

The modern incidence of gang warfare between rival football fans is an expression of the hate between the $(N + 1)$-Self and the $(N + 1)$-anti-Self. Looking for 'reasons' relating to unemployment and family discipline is not an unreasonable thing to do, because those matters relate to the absence of structure which would, if it existed, strengthen the $(N + 1)$-Self – the 'positive' one which fits in with society's needs at the $(N + 2)$-level.

Terrorism and acts of violence against society, or pieces of society, are always performed in the name of some $(N + ?)$-Self, and individual terrorists become fixated at that level: this allows them to lose all contact with $(N - 1)$-Selves and to allow the triumph of the $(N + ?)$-anti-Self. Thus we find the Frenchman, Vaillant, in 1893 claiming to be the 'friend of the people' – and trying to blow up the National Assembly with the cry 'there can be no innocent bourgeois'. He meant to say that there can be no innocent $(N + 2)$-Self ($=$ bourgeois), so the $(N + 2)$-anti-Self must destroy it; but he was wrong to believe that there can be no innocent N-Selves or $(N - 1)$-Selves. But how many young people can resist that rush of blood to the head that comes with a new and powerful awareness of being an (N + 2)-Self? How idealistic is it to see that 'higher' good as transcending the irritations of the 'lower' bad – thus justifying the bad means by the good ends?

What price do we pay for the experience of hate at the (N — 3)-level? The destruction of the structure of (N — 3)-Self must surely produce extensive destruction at all higher levels: so we can lose our friends, our jobs, our loyalties, and (worst of all) our sense of humour?

And at (N — 2)-level we can lose our connectivities, our coherence; even the *mechanism* of the (N — 2)-Self must then be in danger. Cannot this mean physical illness – like the hypertension (which is trying to 'hold the structure together') and anxiety (as a sense of the geometry falling apart) to which we are prone? Is it unreasonable to see this arena as the one where the ulcer and the heart attack begin, where the distortion of structure becomes physical at (N — 2) – the deep cancer of the Self?

The poet Kenneth Patchen, who has suffered all his life with a deep illness, knows and says all this in his lines:

> The animal I wanted
> Couldn't get into the world . . .
> I can hear it crying
> When I sit like this away from life.

6. Of Time and Chance

The idea that events can be 'equally likely' to occur, as when we say that the 'chance' or 'probability' of tossing heads is fifty-fifty, or that some events are 'more likely' than others, seems to be closely related to our notion of p-dimensional traffic on a structured backcloth of available possibilities. For if we glance back to Figure 8 (on page 108) we recall the discussion of 1-dimensional colour-vision trying to cope with a 2-dimensional space of possible events. Such a viewer is incapable of seeing the colour White, since he is only able to see (at most) two simultaneous hues – and to combine them into a 1-simplex (when they become a single colour). Thus, when shown White (which is the 2-simplex ⟨Red, Blue, Green⟩), he cannot decide whether it is Yellow (the 1-simplex ⟨Red, Green⟩), Purple (the 1-simplex ⟨Red, Blue⟩), or Turquoise (the 1-simplex ⟨Blue, Green⟩), because his vision constitutes, we are supposing, 1-dimensional traffic on that 2-dimensional structure (which is what White is). So presumably sometimes he will see Yellow, sometimes Purple, and sometimes Turquoise. And when asked to anticipate which colour is about to appear, consequent upon being shown White, he will say that 'the chances are equal that it is going to be Yellow, Purple or Turquoise'.

But if the viewer's vision were 2-dimensional traffic, in this context, then he would not make that mistake – he would merely say that the colour is White – and there would be no need for him to appeal to the notion of probability or chance. Handicapped as he is with 1-dimensional colour-vision, our viewer can only grope in vain after the 2-dimensional simplex (White): he does it by racing around the edges of that triangle – and the names of those edges are Yellow, Purple and Turquoise.

This experience, when the world of events seems to him to be crazily ambiguous and uncertain, is what he describes by using phrases like 'the chance of the event being Yellow is 1 in 3', or 'the odds against

the event being Green are 2 to 1'. The amusing thing is that the actual
event transcends all three of his events and is unambiguous, certain
and sharply defined as White.

So perhaps we can surmise that *the description of a set of events by
way of assigning them probabilities is a characteristic indication that
the 'seeing-agent'* (or 'process of observation') *is traffic of too low a
dimension on the actual event(s)* – otherwise it would provide us with
that single certain event of which these 'probable' events are only
(geometrical) faces.

The same situation occurs in common gambling events – for example, in throwing a coin or die, or in spinning a roulette wheel. For
the *observed event* is produced by physical circumstances which naturally limit the dimensionality of that event (to one of its faces, in fact).
This process of observation therefore is designed to be low-dimensional
traffic on a high-dimensional event. Thus if I throw a die we normally
agree that we are to look at a single face of it (that which is uppermost)
when it comes to rest – that is to be 'the result of the throw'. But the
die is really a 5-simplex event (it being defined in its totality by six
vertices, which are the faces of the die). Throwing it is a means of
restricting ourselves to seeing any one (but only one) of the 0-simplex
events (one vertex at a time). So the business of throwing the die
constitutes 0-dimensional traffic on a structure of events which is
really 5-dimensional. Throwing only picks out those events which are
0-dimensional, and since there are six of these we experience them,
rather like our handicapped colour-viewer, with uncertainty and
'chance-like' feelings. Which is why we are driven to describe these
0-events (produced by throwing) in terms of probability or chance.
We say that the probability of throwing a 4 is one-sixth (if the die is
'fair' – a tautological way of talking about it). But notice that if the
die is 'loaded', so that it always gives a 6, then the *whole structure*
has been reduced to 0-dimensions; there is now only one event
(vertex) and the 0-traffic (throwing the die) sees it exactly and completely for what it is. No probability ideas are now needed to describe
the possible results of a throw.

But this would also suggest that *when we find ourselves using the
language of probabilities*, in describing our observed events (in any
sphere of observation), then *we have been unconsciously trapped into a
method of observation which constitutes low-dimensional (inadequate)*

traffic on some high-dimensional structure (which constitutes the actual event). And of course this situation is pertinent to hard science, for in physics itself there has been an increasing use of probabilistic language to describe the 'observations' – in modern atomic theory. So, have the physicists drifted into methods of observation (or events) which constitute inadequate traffic on a structure which they therefore cannot hope to observe? Certainly Einstein would have probably thought so – as when he said that 'God does not play dice with the universe'.

When social scientists naïvely adopt statistical (probabilistic) methods to analyse their 'data', in the mistaken belief that they are thereby being 'scientific' in their methodology, then they are unwittingly admitting that the actual events which constitute their universe of observations are of such a high dimensionality as to be beyond the reach of their low-dimensional traffic of observation. But how can they hope thereby to attain an understanding of the actual structure of that universe? Indeed, the methodology based on the use of probabilistic notions is a barrier to the very concept of that structure.

But the situation is actually worse than this, because although we can see some sense in calculations of chances when the events discovered by the traffic are simply faces of a single simplex (even if these chances are inadequate), how can we relate such ideas to a universe whose structure is that of a more general and complicated simplicial complex?

Sequence of events forming a backcloth structure

Even when we are faced with the inevitability of using probabilistic notions to describe events (due to the observation process being of inadequate dimensionality) we are still in difficulties over the interpretation of what we call *likelihood* (of an event occurring). This is because in a sequence of events the observation process can be re-regarded as traffic (of some dimension) on a backcloth structure of events, and this traffic (in any one sequence of observations, or 'trials') effectively moves throughout this backcloth – selecting one event after the other. The sequence does not necessarily exhaust *all* the events in the backcloth, but the latter must contain all the possible sequences which are compatible with the universe of events (what the probabilist would call the 'sample space'). Because of this our sense of 'likelihood' is also traffic on this backcloth – one which anticipates the possible events – and the question is: 'Does the description of events by way of probability numbers adequately represent our sense of likelihood, on that backcloth structure?'

The difficulty is that the calculation of the probability numbers assumes that the events (in the sequence) are dimensionless (in the sense that their dimensions are ignored) and maximally connected, whereas the traffic of likelihood depends very much upon the peculiar way in which the various p-events (events of p-dimensions) are in fact connected. This is because we can only anticipate one p-event following another p-event if they are p-connected in the backcloth 'complex of events'. So if there is to be a distinction between these two kinds of traffic it must depend upon the way the geometry of the complex of events is formed. A simple example can be taken by tossing a coin four times: X_1, X_2, X_3, X_4 denoting these basic trials; then if we denote Heads by 1 and Tails by 0, we get the matrix of the relation between the X's and the sample space of events as shown below. The complex of events will be written as KE(X) and, for example, E_{12} means the simplex $\langle X_1, X_2 \rangle$.

λ	X_1	X_2	X_3	X_4
E_0	0	0	0	0
E_1	1	0	0	0
E_2	0	1	0	0
E_3	0	0	1	0
E_4	0	0	0	1
E_{12}	1	1	0	0
E_{13}	1	0	1	0
E_{14}	1	0	0	1
E_{23}	0	1	1	0
E_{24}	0	1	0	1
E_{34}	0	0	1	1
E_{123}	1	1	1	0
E_{124}	1	1	0	1
E_{134}	1	0	1	1
E_{234}	0	1	1	1
E_{1234}	1	1	1	1

The event E_{1234} means that we threw heads on each trial (of which there are only four), and the bottom row of the matrix shows this by a 1 in each column. This means that, in KE(X), this event is a tetrahedron (3-simplex) and that all the other events are faces of it. So KE(X) consists of one simplex (of dimension 3) only, and this can be expressed by saying that there is maximum connectivity between all the events. In terms of the analysis which finds how many distinct pieces the geometry falls into, at each q-level (compare the discussion of Figure 7, on page 89), we see that the components of KE(X) are as follows:

KE(X):	q-value	components	number
	3	(E_{1234})	1
	2	$(E_{1234}, E_{123}, E_{124}, E_{134}, E_{234})$	1
	1	(all above plus E_{12}, E_{13}, E_{14}, and E_{23}, E_{24}, E_{34})	1
	0	(all the events)	1

The numbers in the right-hand column constitute a vector of numbers, each describing how many distinct pieces the geometry falls into at the appropriate level of dimensionality. Since this gives us a first overall view of the connectivities of the geometry we shall call it the *structure vector* and denote it by $\mathbf{Q}$. It is written, in this case, as

$$\mathbf{Q} = \left\{ \begin{matrix} 3 & 2 & 1 & 0 \\ 1 & 1 & 1 & 1 \end{matrix} \right\}$$

where the small numbers at the top show the dimensional value corresponding to the number at the bottom. Such a string of 1's is typical of a complex which is only a single simplex, with its maximum connectivity.

We can now finally arrive at the answer to our problem, based on this particular structure of KE(X). For the number of 1-events is the number of events with labels like E_1, the 2-events are things like E_{12}, the 3-events like E_{123}, and the 4-event is E_{1234}. So the probabilist who says that the events are 'equally likely' first says that there is a total of 16 possible events (counting the additional event represented by the first row, which is that in which not a single one of the X's occurs (all failures)) and so he would say

prob(any 0-event) = 4/16, prob(any 1-event) = 6/16
prob(any 2-event) = 4/16, prob(and 3-event) = 1/16.

If we do not attribute equal chances to all events (so, at the N-level we would not say that each probability is of value 1/2) then the numbers become different. But they are calculable by attaching new fractions to the X's – the details of which are not worth pursuing in this book. The general result, telling us how the events have their chances distributed throughout the whole complex, is called the *binomial distribution* by mathematicians.

This indicates that the notion of probability (which ignores dimension) is associated with a certain *traffic* on the structure KE(X); that this traffic can be represented by certain numbers (probabilities) associated with the simplexes (which are the events) of the geometry of KE(X); and that this is traffic on an underlying structure of events, that is, KE(X). It is a fact that the theory of probability distributions rests on this foundation – that the underlying geometry is that of a single simplex whose structure vector **Q** is a string of 1's, as above. And it is also highly plausible in the case of the throwing of dice or coins etc. – as our ancestor gamblers of the seventeenth century soon made it their business to find out.

But what happens to this traffic if it is forced to move on some structure KE(X) which looks like that of Figure 7 (page 89)? There we have seen that the structure vector is given by

$$\mathbf{Q} = \begin{pmatrix} 3 & 2 & 1 & 0 \\ 2 & 6 & 3 & 1 \end{pmatrix}$$

which means that the structure of events is a long way from being a single simplex. Or what if we examine some of the structures shown by Mulhall's patient, John (see page 98), where again the structure vectors were not often strings of 1's? Indeed, in Week 1 of the treatment we find that there is a structure KE(X) – showing how the husband's actions are connected – with a vector given by

$$\mathbf{Q} = \begin{pmatrix} 2 & 1 & 0 \\ 5 & 2 & 1 \end{pmatrix}$$

Such underlying structures can clearly carry traffic which corresponds to the happening of events: so the notions of chance or probability can be attached to such structures (indeed they must be so attached if they refer to events which depend on the vertex set – genuine traffic), and the connectivities presumably affect what that can mean.

G

Chance is unlikely

The gambler works on the assumption that his intuitive idea of *likelihood* is reasonably well represented by the probabilist's definition of chance, and the theory of probability distributions (however much it is dressed up) works pretty well in those contexts to which the gambler applies it. This, I would claim, is *because the underlying structure of the events* he is dealing with *is that of a single simplex*, with its maximum connectivity properties exemplified by the simple structure vector (of a string of 1's).

The point about the connectivity is that it affects the flow of traffic throughout the geometry: 1-traffic needs 1-dimensional connections before it can move from one piece of geometry to another piece, and so on. If the traffic which represents the probabilities of (say) p-events is also to represent our sense of the *likelihood* of those events occurring then it must do so both before and after any actual happening. But this means that, *before* we test the trial, there is a certain distribution of probabilities (numbers) all over the complex KE(X) and that, *after* we let the trial go ahead, these numbers must have all shifted around so as to pick out that one event which has now in fact occurred. This shifting around corresponds to the movement of the probability traffic over KE(X), and it picks out the actual event by a representation in which the probability of *that* event is now 1 (for it is now certain) whilst the probabilities of all the other events become 0. But the traffic cannot move around the geometry unless that is adequately connected at all dimensional levels: this traffic (which is the probabilist's traffic) always needs geometry with a Q of the simple kind.

On the other hand our sense of likelihood is heavily dependent on our sense of the underlying structure of KE(X). And in most cases we would expect this to be more severely warped and disconnected than is that of a simplex. So then the probabilist's distribution theory about chances of events happening is unlikely to correspond very closely with that traffic which we experience as likelihood. So we even have a sort of *criterion* for when it is fair and reasonable to use probabilistic statistical theory on our data – it is *when the structure vector of the complex of events is a string of 1's*, and not otherwise.

It is perhaps ironic that this situation troubles the probabilists themselves on ground which is very close to home. For if we look closely at the matrix above, and contemplate the conjugate complex KX(E) – so we look down the columns instead of along the rows – then we can see that the X's (as simplexes) always share exactly four vertices (selected out of the rows) pairwise. Thus we see that X_1 and X_2 share the vertices E_{12}, E_{123}, E_{124}, and E_{1234}. So when we begin to list the things which are q-connected, for any q-value greater than 3, we find that the X's are always in distinct components. In fact the structure vector for KX(E) is

$$KX(E): Q = \begin{pmatrix} 7 & 6 & 5 & 4 & 3 & 2 & 1 & 0 \\ 4 & 4 & 4 & 4 & 1 & 1 & 1 & 1 \end{pmatrix}$$

so even the usual 'sample space' does not meet the requirements of our criterion, in the conjugate complex! But does that matter?

Yes it does, because the conjugate complex is the underlying structure which is appealed to when probabilists apply what is known as *Bayes' Theorem* to problems. This is a theorem which tells them how to turn the probabilities inside out, and to find probability values for things like the X's – given the set E as a notional set of 'elementary events'. And there have always been difficulties with Bayes' Theorem, as probabilists have found it not easy to apply and sometimes 'unreliable'. From our discussion we would expect it to be reliable if the events to be examined were not greater than 3-events (dimensionally) – but this will be difficult if the X's are as shown, since each is a 7-event. So it looks as if Bayes' Theorem (which has been used by many people in the applications of probability distribution theory in medical diagnosis situations!) should be re-examined in the light of this structural approach.

The reader might wish to bear in mind the fact that when his doctor talks about his 'chances of recovery' or of the 'chance of his contracting some specific disease', he is treating his patient as a piece of geometry which is only a face of some single simplex and, in addition, ignoring his dimensions.

Sense of Time derived from a sequence of Events

In discussing the idea of dimension and how we have inherited it from the ancient Greeks we have appealed directly to the intuitive notion of a relation between things: in the Euclidean space E^3 those 'things' are points, lines, etc. And this idea is more in the tradition of Leibniz than in that of Newton – although it certainly did not originate from the former. The idea that (ordinary) 'space' is composed of the relation between objects is also of ancient Greek origin and is to be found in Aristotelian philosophy. The alternative view, which is widely assumed among technological scientists, is one which became important with Descartes and, subsequently, an assumed foundation of Newtonian science. This latter view amounts to the absolutist position expressed by saying that 'space' is given *a priori* and that 'objects' are things which sit in it. This absolute idea about the 'given' space, once expressed mathematically, becomes the stumbling block to extending our notions of dimension – because it places the burden on our intellects to imagine higher dimensions which transcend those of the mathematically defined E^3. It means also that the Newtonian space is a sort of Euclidean waste-paper basket where other 'things' (called objects) can be thrown. Such a space is naturally rigid and unwarpable, so it is not surprising that the Einstein theory of this century came as such an intellectual shock to many scientists – from which some have never recovered to this day. That rigidity extends also to the notion of dimensions itself and induces in us the obsession with a 3-dimensional space.

It was presumably inevitable that the Newtonian view of space should also have been accompanied by a Newtonian idea of 'time' – and one based on the same sort of absolutism. So, in the Newtonian sense 'time' becomes a sort of temporal waste-paper basket, sitting there and waiting for 'events' to exist in it. Such a view must exclude, as 'unscientific', the possibility of a subjective time as part of the fundamental fabric of the 'world of events'.

But it is clear that in this work we must pursue the idea that *time is the manifestation of relations between events* – comparable to our view of (general) space. This view means that the events are *a priori*, in our experience, just like objects are in our experience of space. And then we must reject the Newtonian idea that time is like some ever-rolling stream of vacuous 'moments' just waiting to be filled by 'events'. In the words of Aristotle, 'time is the measure of change with respect to before and after', and to this we must add that the 'change' refers to the experience of some *structure of events*. But the idea of change, for its own sake, is not adequate for the expression of time. What we need also is what the mathematicians call a *total order* of the events. That is to say, if we think of the possible events as comprising a set E, then we must be able to contemplate them as arrangeable in a sequence

$$E_1 \rightarrow E_2 \rightarrow E_3 \rightarrow \qquad \rightarrow E_r \rightarrow E_{r+1}$$

and this sequence expresses the notion of 'before' and 'after'. Nor do we need, at this stage, to insist that we all agree about that sequence – it is not to be an absolute total ordering of the members of the set E.

Of course, this idea is not hostile to the Newtonian sense of time. It is too fundamental for that, and the 'absolute time' fits such a scheme without any trouble. But I shall argue that another *structural* sense of time also fits it admirably and, in addition, is more relevant to our actual experience of time.

For we are told that time is frequently 'an enemy', and that it often has different 'speeds' (whatever that can mean in orthodox terms). Often we know that 'time flies', and people complain that 'time drags'. And although it is commonly an enemy we also believe that it is a 'great healer'. Then there are questions about time which we often ask but which Newtonian time is not equipped to answer. Children often want to know 'how long will it take me to grow up?', or 'how long will it be before I get better?'. To answer either of these in a truly 'scientific' manner is ludicrous (for example, let the answer be 15 yrs 8m. 10 d. 7 hrs 15 min. 11 sec., to the first question), so it is asserted, or assumed, that the question is not scientific: or that 'at our present stage of scientific knowledge we cannot give a precise answer'. But children are not the only culprits in this context. Why should not a businessman ask the question 'how long will it take for my new business to be a success', or the (N + 3)-Self ask the question 'how long will the war last?', or the lover ask 'how long will you love me?' or 'when did you stop loving me?' (answer: at 15.32 hours last Tuesday), or the artist ask 'how long will this work of art endure?'.

Although 'science' cannot answer these questions in terms of its clock time, its assumptions about that sort of measured time, and its use of it, already contain strong hints about the way out. The work of Einstein produced the first big shock to technical clock-time thinking, although most scientists get along in their old ways without letting that worry them. We can see that in the discussion that follows.

Scientific clock-time

For the past 250 years practising scientists have been getting on quite well with Newtonian clock-time, and have steadily imposed it on us in an increasingly technological world, because they have required little of it in the way of philosophical justification. All the physicist has asked is that it should fit into his theoretical discussions as easily as do other quantities. By 'fit in' he means only that there shall be, in some standard manner, a mathematical way of representing the thing called 'time', and that this mathematical way shall be no more complicated than the way which he uses to talk about space. This has meant, for such a theoretician, that it has been quite sufficient for him to represent time by a *time-axis*. With such an axis, one starts at a convenient origin on a line and extends outwards for ever – the line being marked with an arrow at the far end, where a small letter t stands beside it. Then the scientist has an algebraic symbol, his t, which he can play around with as he does his x, y and z letters (for space-axes). His vague idea that time is an ever-rolling stream is adequately represented in his mind by this variable, t, progressing through the continuum of the real numbers (which happily possess the property of total ordering mentioned before). The idea of the progression through these values becomes modelled as an imaginary 'point' which is allowed to move along this axis. Thus he has a model of Newton's own description of time: 'Absolute, true and mathematical time, of itself, and by its own nature, flows uniformly on, without regard to anything external. It is also called duration.'

With this theoretical model in his mind the technologist has pursued, with ingenuity and success, the practical problem of finding various horological devices which illustrate it. But none of these can ever be said to measure the Newtonian absolute time, because such a notion must transcend the measuring of any time – assuming as it does that there is some 'great pendulum in the sky' which is beating out the seconds *correctly*. Thus a grandfather clock seems to have a certain regularity about it, as it cycles through its repetitive conditions, and this intuitive sense of its regularity is what we must rely on as an expression of our in-built sense of Newtonian time. It is quite ideal to ask 'how regular is the grandfather clock?'. For this cannot be answered without referring to another kind of clock, and so on.

This model of Newtonian time can be represented by the simple geometry of Figure 21, where a distinction is made between the 'moments' (often called the now-points) and the 'intervals' (or durations between successive now-points). The arrow indicates the difference between past and future, relative to any particular now-point.

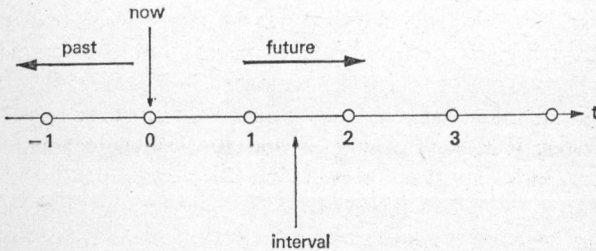

Fig. 21   Newtonian time structure

But what we notice immediately about this geometry is that it is a particularly simple *simplicial complex*, with an unending set of vertices (the now-points) which are the end-points of a sequence of edges (1-simplexes) which are the time-intervals. If the now-points are the *events* which we can experience, then they correspond to 0-simplexes, whilst the time-intervals (the 1-simplexes) identify a *dimensionally different* set of events.

So the Newtonian idea of time is already an expression of a simple view of the world of 'events' – one in which there are two kinds of events and these are dimensionally different. Naturally we can refer to them as *0-events* and *1-events*. Then the orthodox technologist's view of time is one which insists that any 'scientific event' must fall into this particular structure of the time model. This means that observable events have to be things which we can associate with this geometry of Figure 21, and if it stands there (as the temporal waste-paper basket) waiting for the events to be associated with its parts, then it is acting as a *static backcloth* whose geometrical connectivities tell us how events are related to each other. Any such set of events will therefore constitute *traffic* on the backcloth: 0-events constituting 0-traffic and 1-events constituting 1-traffic.

It is because of this kind of assumed backcloth that we can view the Newtonian clock-time as a peculiarly *linear* concept. The linearity of the complex of Figure 21 is precisely the 1-dimensional property which characterizes it. Hence any scientifically observable set of events must have this linearity imposed on it by the model, and it must also have the dimensional limitations imposed on it. Perhaps it is not surprising that this model eventually became untenable, at the deeper levels of physics, and the Einstein theory (which warped the geometry of space) effectively altered this backcloth geometry of events.

Einstein's relativity theories

In his theories of relativity, the Special Theory of 1905 and the General Theory of 1915, Einstein found it necessary to break from the orthodox assumption that the time-axis could be treated as independent of the space-axes. (This already hinted at the difficulties inherent in the absolutism of the spatial and temporal waste-paper baskets.) The difficulties had arisen through measurements of the velocity of light – which turned out always to give a fixed constant value (usually denoted by the letter c, with a value of 186,000 miles per second) which was independent of the velocity of the source or of the observer. This breakdown of the independence (which amounted to saying that space measurements and time measurements had to be regarded as inextricably mixed up) was expressed by postulating the so-called space-time continuum – a concept which was essentially mathematical in its nature and which combined the traditional 3-dimensions of geometrical space with the observer's 1-dimensional sense of time. (It would be quite inappropriate to delve into the mathematical details in this book, but the interested reader can find a discussion of the theories in the references in the bibliography.) This combination of geometrical and time axes, particularly in the General Theory of Relativity, had to be arranged in such a way that although for one possible observer in the universe the time and space might be quite separate and independent, yet for another observer this would be quite impossible (and 'laws of nature' would have to be statements which took all this into account). Hence the time-pattern which any particular observer (a pseudonym for an experimental physicist) will find is itself dependent upon (that is to say, is a function of) space-like properties of the system wherein the observer is situated. Thus the 'new' time-intervals will be a function of velocity and position of the body being observed. This strange mix-up cannot be unravelled for all observers at all positions (in Newtonian space) and all moments of (Newtonian) time. Indeed it becomes impossible to speak of a 'moment of time' as if it were something that could be simultaneously observed and understood by different observers. Thus the idea of Newtonian absolutism as well as the idea of independence (of time and space) had to be eliminated from the theoretical structure.

But it is perhaps ironic to notice that this was only achieved by replacing one kind of absolutism by another. In place of the idea of absolute time and absolute space the theory of relativity introduced the idea of an absolute *signal velocity* – where the word 'signal' refers to the *light signal* which acts as the messenger bringing information about the position of the observed body. This is because, in a practical situation, it had to be assumed that all the information about the geometry of the system is carried by the light signal. If this signal has a finite velocity (which it has, and which is the value denoted by c) then it builds into all the observations its own sense of delay. The absolutism of this philosophy is to be found in the observation that however it is measured it turns out to have the same constant value, c. This in turn leads to a mathematical 'proof' that no body 'can travel faster than the speed of light' – which the following discussion shows to be a tautology, and scientifically trivial.

Suppose that I am a businessman living in an ancient town when communications were more primitive than they are today. Suppose a fellow businessman comes to see me at my house and rides away on horseback. After he has gone I decide that it is important to obtain another message from him, and so I send out my light-signal – in the shape of a fast runner or clerk. This runner runs at a standard speed denoted by c, and he races after my acquaintance. Now if the latter rides at a speed less than c my clerk will catch him, exchange messages, and bring the word back to me. But if my new acquaintance travels at a speed which is greater than c, then my clerk will never catch him. He is under instructions to travel for a year and a day, and then to return. So it is possible that my clerk will return after two years and two days and announce that the man does not exist – nor does he, as far as my experimental physics is concerned. As far as doing business with him is concerned I must accept this and assume that the man has literally vanished.

In this kind of medieval business world it is obvious that the whole thing (of 'events' – that is to say, the world of making contact with objects by using a suitable signal for detection purposes) is held together by the clerks who travel about with velocity c. If anything travels faster than c then the clerk will never find it. So it is not that a body 'cannot' travel faster than light, but only that *if we detect things by the light signal, then we shall never observe such a phenomenon.*

The Einstein theory effectively says that 'velocity' comes before notions of 'position' or 'time' (in the Newtonian sense), and that since velocity is associated with (in the Newtonian sense) *both* a position *and* a time, it must be an *event* which requires both of them before it is recognized. But this means that 'velocity' will then be a 1-simplex, denoted by $\langle X, T \rangle$, where the X refers to the Newtonian x-axis of some space and the T refers to the time-axis (of Figure 21). But this means that we can get a simplified version of the Einstein theory of relativity by regarding Figure 22 as representing the new space-time structure. It is really the new time-structure, if we adopt our Aristotelian–Leibniz point of view and say that the events (which are now 1-simplexes) determine the sense of time (and not vice versa).

In this geometrical representation the set of possible events, E, are connected to give a new simplicial complex (refer back to Figure 21) which is made up of an unending series of edges (1-simplexes) which have become the now-edges (see the now-points of Figure 21), and the time-intervals have become the pieces of area (the appropriate 2-simplexes) which join one such now-edge with another now-edge. So Einstein was effectively saying (he did not express it in these terms) that when we measure a sequence of events in the world we must associate them with a new backcloth time-structure (Figure 22 is a simple version) in which the traffic which we call 'time' is identified as follows: the now-events are 1-simplexes (1-traffic), and the time-intervals are 2-simplexes (2-traffic) on that backcloth.

This should be compared with the orthodox Newtonian time-structure, whence we notice that Einstein's theory, although still an absolutist one, has altered the *dimensions* of the traffic we call 'time'. It has been increased from being a 0-traffic cum 1-traffic to being a 1-traffic cum 2-traffic. Not surprisingly we should expect the complete

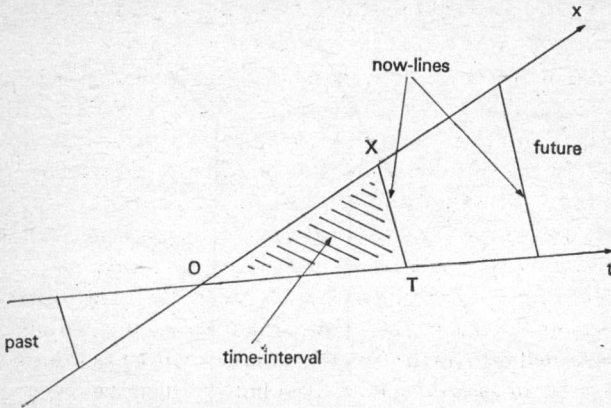

Fig. 22   Simplified Einstein time structure

Einstein theory to replace the single x-axis measurement (denoted by X in Figure 22) by the normal three spatial axes of x, y and z. This situation is denoted in Figure 23 below, and there we see that the General Theory requires us to identify the *dimensions* of the time traffic in the following way: the now-events are 3-simplexes (3-traffic), and the time-intervals are 4-simplexes (4-traffic) on the backcloth.

Popular discussions of Einstein's relativity theory often concentrated on the novel idea of a 'fourth dimension' in physics. This was usually identified with Newtonian time in an obvious way, as being a fourth quantity like x, y or z and equally independent as each of them is. But this completely overlooks the significance of the dimensions of Figure 23. Furthermore, it consisted in assuming that the orthodox Newtonian time could be transferred to relativity time by the simple expedient of multiplying by a factor c (the velocity of light).

How old are you?

Orthodox or Newtonian time plays a strangely dominant role in our official lives. It means that every time the Earth completes an orbit around the Sun the government (and other lackeys of the social

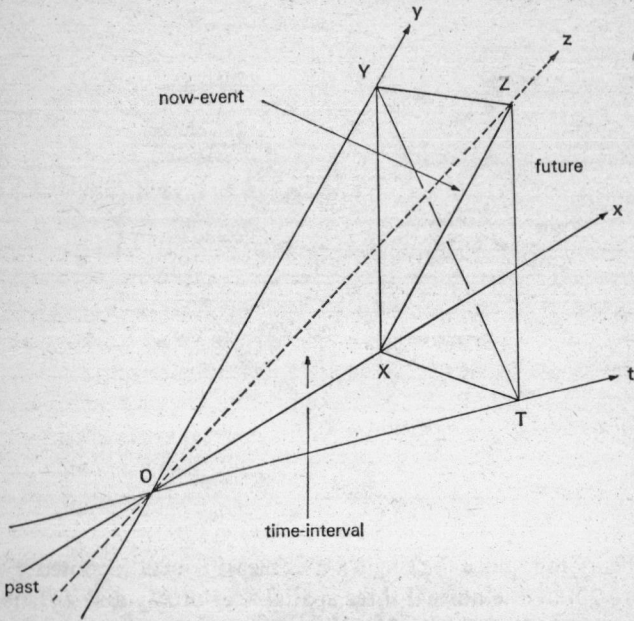

Fig. 23 General Einstein time structure

orthodoxy) adds the number one to your age. So your progress through life becomes mapped into the structure of Figure 21, with the now-points being numbered 0,1,2, etc. This produces a *linear view* of your life and this means that all the structure which you are patiently building up (to form your character, your body, your mind, your social and personal relationships, your work, etc.) becomes condensed into a point (a 0-simplex) and is then represented completely by this number which the government calls your 'age'.

But when they say that you are 56 years of age, you know very well that there is a real sense in which you are also 46 and 36 and 26 and 16 and 6. You are all these 'ages' at once – because you are all the structure (of a large p-simplex attached to many different complexes, which go to make up your life and experience) which has been slowly built up through each of these fanciful years. So it would seem that when you are 'behaving like a child' it is because you *are* a child (and why not?) and now that 'you have become a man' it is *impossible* to 'put away childish things'. To do so, effectively, is to destroy an important part of your own structure-as-a-man: what kind of crazy living is that? Then, when life begins to decay and all the big structure begins to crumble, what more natural than that you should be left with your earlier childish geometry – your 'second childhood'? The rapport between grandparents and grandchildren contains this common element of identified structure which they both can share: the children because they know nothing else, the adults because they do. Being disconnected from your childhood can mean being disconnected from your childhood structure – so your roots have been destroyed – and you can only experience all the *empty spaces of anxiety*.

When you reach middle age, and a lot of the fury and struggle of building your 'life' (which means your geometry!) is over, then you are amenable to awakening again to the geometry which came more easily and instinctively in the sweeter years of youth. So a middle-aged man will fall in love with a younger woman – because his younger geometry is searching for the younger geometry of his (Jungian) anima self? That geometry is still there and still very real; it is no mean fantasy – although the social norms might want to insist that it is (because these norms believe in the linear idea of progression through those 'years' – what is behind is gone and lost, and only the 'point' of the present is now available). And so we can find the 'ageing' woman who is infatuated with a young man – because she is living again through the geometry which she built up at the age of 25, and *which is still there* as part of her structure (and why not?). If there is tragedy or farce in this situation it is surely because the one partner is turning back to an earlier geometry while the other has yet to build some geometry which is not yet there. The scornful cry to 'be your age', under these circumstances, can only be answered by 'what age am I?'. Is it wrong to be 16 and 46? Are there not circumstances in which it could be more important to be 16 than 46, and sometimes the other way round?

Is not your 'age' really some p-dimensional structure, some p-simplex in a well-defined complex which has grown and contracted throughout the living process? Should not our sense of 'time' some-how express this experience of working through big-dimensional structures, and has not hard science already anticipated the essence of this view?

Multidimensional time

It is not an exaggeration to say that Einstein's Relativity Theory is essentially a theory of time. It demands that the classical Newtonian time traffic (on the geometry of Figure 21) be replaced by a new and higher-dimensional traffic on a new structure which is dimensionally equipped to carry it. It is reasonable therefore to see this scientific advance as strong evidence, in orthodox scientific circles, for time to be regarded as a *multidimensional traffic* on a suitable structure of events – which must therefore also be regarded as being so related to each other as to provide us with a *complex of events* of suitable dimensions. But we know also from our discussions that the set of events should be hierarchically organized and that we must then distinguish between the events at N-level, (N + 1)-level, and so on. Then at any one of these levels we would expect the events to generate a structure, say S(N), with a representation in a space E^n (where n is suitably large).

Now we will have to say that, with respect to the backcloth S(N), there will be a time experience (in which we are aware of the total ordering of the events) which will be graded into many dimensionally different orderings. This will be because the structure S(N) will normally contain

0-simplexes (where an 'event' is a point),
1-simplexes (where an 'event' is a pair of points (edge)),
2-simplexes (where an 'event' is a triple of points) etc.,
so that generally we get
p-simplexes (where an 'event' is a set of (p+1) points).

If an 'event' in this structure S(N) needs to be represented by a p-simplex we might as well call it a *p-event*, and this will correspond to a p-dimensional now-event. But in that case the corresponding time-interval will have to be a (p + 1) simplex (because such things will join up successive p-events). And this situation will face us throughout the whole structure S(N) of all possible events. The difference between a 1-simplex as a 1-event (a now-event) and a 1-simplex as a 1-interval (a time interval between two successive now-points or 0-events) is that in the latter we can *order* the end-points of the 1-simplex, whilst in the former we cannot do any such thing. So when we think of a now-event as a p-event we are saying that the $(p + 1)$ points which go to make up this p-simplex cannot be ordered (in the linear Newtonian sense of Figure 21) – they must be accepted as a new unit (a new now-event) and the associated time-intervals are then some experienced time traffic which allows us to 'move' from one such p-event to the next such p-event.

No doubt some things, like physics-type particles, can often be associated with 0-events and 1-intervals, as is done in orthodox technology and has been since the time of Newton. But most of our experiences are not like this – certainly not in the world of human relations, at any of the levels $(N + 3)$ down to $(N - 4)$. Newton said that time is experienced as events move from one point to another point: Einstein said that time is experienced as events move from one tetrahedron to another. This thesis is saying that time is a multi-dimensional experience as events move from one set of simplexes to another such set, and in that set there will at any one occasion be events at various dimensional levels. Thus we shall experience 0-events, 1-events, 2-events, 3-events, ... p-events, ...; the 0-events will move from one point to another, the 1-events will move from one edge to another ..., the p-events will move from one p-simplex to another.

If we try to pretend that each of these experiences is Newtonian, so that even a p-event is really a 0-event, then we shall effectively be treating the geometry S(N) as if it were that of Figure 21. This means that we shall be trying to warp the geometry of the events and experiencing t-forces of repulsion in the process. So forcing our time experience into the clock-time mould will be to subject ourselves to structural t-force stresses – perhaps the most striking of which is the sense of *impatience*?

Some speculative sums

Now a p-event is represented in a p-dimensional space by a polyhedron with $(p + 1)$ vertices: Figure 24 shows two successive 2-events, the first is the simplex $\langle A_1 B_1 C_1 \rangle$ and the second is $\langle A_2 B_2 C_2 \rangle$.

The connecting lines join up the vertex sets of these two 2-events, so the associated time-interval is a 3-interval – since it requires a 3-dimensional space for the representation of something like the simplex $\langle A_1A_2B_2C_2 \rangle$.

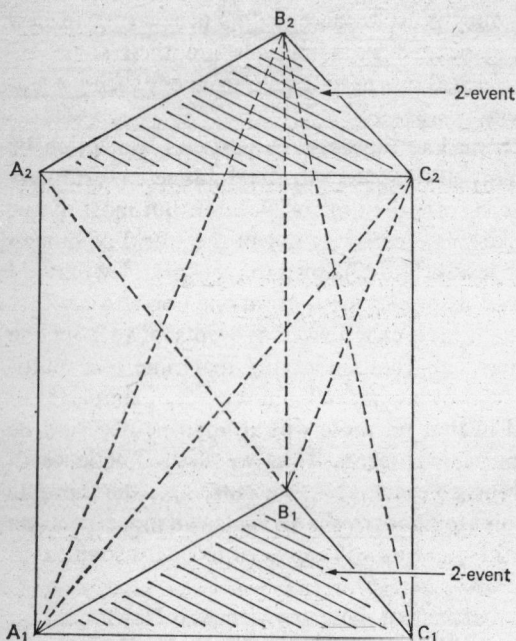

Fig. 24   Successive 2-events with connecting edges

We can now speculate (and without some serious research into the problem it can be no more than speculation at the moment) as to how such higher-dimensional p-events and $(p + 1)$-intervals might be experienced in terms of standard clock-time. That sort of time is shown in Figure 21 and the intervals are entirely defined by edges (1-simplexes). So in the first place we would realize that a p-event would have to be described by a Newtonian clock-time observer entirely in terms of its edges (the lines which join pairs of vertices). The number of such edges will be the number $p(p + 1)/2$. In the case of a 2-event, $p = 2$ and so this number is 3 (and a triangle has 3 edges). If that is all that the observer can handle then, in terms of the clock-time unit (1 second, 1 hour, etc.), the p-*event would need at least* p(p + 1)/2 *units* to establish itself as an actual occurrence. We can say 'at least', because it would be manifest by the observer going around the edges in some sort of order, and it is not clear how he would decide on that order or that he might have to go over some edges twice in order to completely traverse the edges of the polyhedron.

We can refer to this speculative clock-time as the *consolidation time* for the p-event, given the vertices of that event: the consolidation time is then the time it takes to make the jumble of vertices into a polyhedron. This gives us a possible table to illustrate the numbers involved.

Event	*Least units of clock-time needed for consolidation at N-level*
0-event	0
1-event	1
2-event	3
3-event	6
7-event	28

We shall regard this consolidation time formula as being the sort of number which an N-level observer might make of an N-level p-event: so it is hierarchically bound. This is because we would need to take a bigger view of the possible events if we allowed the observer to move up the hierarchy to the (N + 1)-level. For from that higher level the observer would be able to see *all the faces* of the polyhedron (see all the sub-events of the p-event) and he would need a clock-time for each of them. So in the example of a 2-event, as in Figure 24, the (N + 1)-observer would see all the events: three 1-events (the separate sides of the triangle) plus one 2-event (the sides of the whole triangle). Although this appears to count the edges twice it is necessary for that to happen if the observer is to really distinguish between the dimensions of the sub-events.

This now gives us a larger consolidation time, which turns out to be given by the formula p(p + 1). 2^{p-2}. Notice that when p = 2 this gives 2.3.1 which is 6. So now we list a short illustrative table of such a *superconsolidation time*.

Event	Least superconsolidation time
0-event	0
1-event	1
2-event	6
3-event	24
7-event	1792

In either of these two cases we notice that events which are genuinely 0-events or 1-events have consolidation times which agree with the expected clock-times. But we notice that whereas the consolidation time (hierarchically bound) for a 7-event might be 28 days, the corresponding superconsolidation time will be approximately 5 years.

If we turn our attention to the genuine $(p + 1)$-intervals associated with successive p-events we can find similar formulae for the number of clock-time edges needed to move from one p-event (a set of $(p + 1)$ points) to another p-event. We can take first the simple hierarchically bound view in which we see only the edges (the dotted lines shown in Figure 24) and count them only once; we then add to that number the consolidation time, to get a final interval between the 'end' of one p-event and the end of the next. This gives us a total time (number of edges) of $(p + 1)^2 + p(p + 1)/2$ units and the following table:

Event	Hierarchically bound time-interval
0-event	1
1-event	5
2-event	12
3-event	22
7-event	92

If we then take the $(N + 1)$-level view we need to move from the first p-event (from each of its vertices) to each *subset* of the vertices of the second p-event; adding these total edges together we then add the previous consolidation time. This process gives us the formula: $(p + 1)^2 . 2^p + p(p + 1) . 2^{p-2}$ which equals $2^{p-2}(p + 1) (5p + 4)$. In Figure 24 this process would mean that the observer had to count the edges A_1A_2, A_1B_2, A_1C_2 and then the *pairs* of edges $\{A_1A_2, A_1B_2\}$, $\{A_1A_2, A_1C_2\}$, $\{A_1B_2, A_1C_2\}$ and then triples of edges $\{A_1A_2, A_1B_2, A_1C_2\}$, and then repeat this starting with B_1 and C_1 in turn – and then add in the superconsolidation edge-count for the upper 2-event.

Now we get a typical table of hierarchically unbound time-intervals:

Event	Hierarchically unbound time-intervals
0-event	1
1-event	9
2-event	42
3-event	152
7-event	9984

So if it takes 1 day for one 0-event to be succeeded by the next 0-event, it will take (as seen by the next-level observer) 9 days for successive 1-events and 152 days for successive 3-events.

Is it now too fanciful to suppose that, even if we do not take the formulae too seriously, these interval times might be also an indication of *survival* or *duration* times – for 'events' which are associated with a multidimensional structure? Are they not suggestive of the endurance of such things as works of art, or of science? How long will a Beethoven symphony 'endure' relative to a pop song? Is the difference determined by the dimensionality of the two works, in some such way as here indicated? Why do we still admire ancient works of art – are they not 'timeless' in the sense that they represent p-events of such magnitude that their endurance time-intervals have not yet run out? If the artistic treasures recovered from the tomb of Tutankhamun (around 1350 B.C.) still speak to us with a vivid and lively tongue, is it not because they are large p-events as works of art and science, and that we are (in our culture and times) still inside that duration interval which corresponds to their 'life'? The life span of an individual, or of an institution, or of an empire (various collective Selves) will be dependent on the dimensionalities of those p-events which they represent. So 'growing up' is a building of the p-event, and dying is a disassembling of it – into lower-level hierarchical pieces (points) which lose their coherence (as event-units) and become lower-level points in the total hierarchy of events.

Even though we should regard the formulae given above as rather crude, hiding as they do many assumptions about the relation of clock-time to higher-dimensional time, yet they give us some idea of the scales of the magnitudes involved. Thus, if we take the super hierarchical view of p-events, we notice that a 7-event needs a factor of the order of 10^4 clock-time units for its associated time-interval. If we evaluate the formula for the 9-interval associated with successive 8-events we similarly find that the factor is close to 25,000. These calculations give results which are very reminiscent of others given by P. D. Ouspensky (who was a leading disciple of the teacher Gurdjieff) in his book *In Search of the Miraculous*. There he presents a table of 'times of cosmoses', which seems to correspond to our notion of hierarchical structures – the 'times' referring to the 'natural life-span' of the specified world. The crucial times for the microcosmos (by which he means Man) he associated with those of 'quickest impression', 'breath', 'waking and sleeping', and 'life' – and of course the figures are approximate, indicating chiefly the orders of magnitude, but are based on some commonsense biological information. Ouspensky's times are as follows:

quickest impression	*breath*	*waking/sleeping*	*life*
10^{-4} secs	3 secs	24 hours	80 years

These compare with the following list, assuming hierarchically unbound time-intervals of successive 8-events at successive levels (and taking 24 hours as a given fact):

1.34×10^{-4} secs	3.4 secs	24 hours	70 years

I think I prefer the three score years and ten to Ouspensky's 80!

References

1. HOGBEN, L., *Science in Authority*, Allen and Unwin, 1963.
2. KOESTLER, A., *The Ghost in the Machine*, Pan Books, 1970.
3. LEECH, G., *Semantics*, Penguin Books Ltd, 1974.
4. ATKIN, R. H., 'Multidimensional Structure in the Game of Chess', *Int. J. Man-Machine Studies*, *4*, pp. 341–62, 1972.
5. ATKIN, R. H., HARTSTON, W. R. and WITTEN, I. H., 'Fred CHAMP, positional-chess analyst', *Int. J. Man-Machine Studies*, *8*, pp. 517–29, 1976.
6. CROMBIE, A. C., *Augustine to Galileo*, vol. II, William Heinemann Ltd, 1959.
7. PEDERSEN, O., and PIHL, M., *Early Physics and Astronomy*, MacDonald and Janes, London; American Elsevier Inc., New York, 1974.
8. MACH, ERNST, *The Science of Mechanics*, Open Court, 1960.
9. KUHN, T. S., *The Structure of Scientific Revolutions*, University of Chicago Press, 1962.
10. ASIMOV, I., *Foundation, Foundation and Empire, Second Foundation*, Panther Books, 1960 et seq.
11. HELLER, J., *Catch-22*, Corgi Books, 1974.
12. HASEK, J., *The Good Soldier Schweik*, Penguin Books Ltd, 1951.
13. LEACOCK, S., *Literary Lapses*, The Bodley Head, 1911; Penguin Books Ltd, 1944.
14. POTTER, S., *Lifemanship*, Penguin Books Ltd, 1962.
15. SOLZHENITSYN, A., *One Day in the Life of Ivan Denisovich*, Sphere Books, 1970.
16. BROWN, DEE, *Bury My Heart at Wounded Knee*, Pan Books, 1972.
17. HUXLEY, A., *The Doors of Perception*, Penguin Books Ltd, 1971.
18. OAKLEY, ANN, *Sex, Gender and Society*, Maurice Temple Smith Ltd, 1972.
19. PIAGET, J., *Structuralism*, Routledge and Kegan Paul, London, 1971.
20. ATKIN, R. H., *Combinatorial Connectivities in Social Systems*, Birkhauser, Basle, Switzerland, 1977.
21. DUNNE, J. W., *The Serial Universe*, Faber and Faber, 1934.

Bibliography

ATKIN, R. H., *Mathematical Structure in Human Affairs*, Heinemann Educational Books Ltd, 1974.

ATKIN, R. H., 'Methodology of Q-Analysis – A Study of East Anglia I', Research Report V, University of Essex, 1975.

ATKIN, R. H., 'Q-Analysis, Theory and Practice', Research Report X, University of Essex, 1977.

BARKER, DIANA, and ALLEN, SHEILA, *Dependence and Exploitation in Work and Marriage*, Longman, London, 1976.

BOYD-HIGGINS, M., *Wilhelm Reich: Selected Writings*, Farrar, Strauss and Cudahy, 1960.

EDDINGTON, A., *Space, Time and Gravitation*, Cambridge University Press, 1921.

VON FRANZ, M.-L., and HILLMAN, J., *Lectures on Jung's Typology*, Spring Publications, Zurich, 1971.

FREEMAN, E., and SELLARS, W., *The Philosophy of Time*, Open Court, 1971.

JUNG, C. G., *Man and his Symbols*, Aldus Books Ltd, 1964.

LANCZOS, C., *The Einstein Decade*, Elek Science, London, 1974.

LUCE, GAY G., *Body Time*, Maurice Temple Smith Ltd, 1972; Paladin, 1973.

LUCIE-SMITH, E., *Eroticism in Western Art*, Thames and Hudson, 1972.

DE QUINCEY, T., *Confessions of an English Opium Eater*, The London Magazine, 1821; New American Library, 1966.

RICHARDSON, J., *Georges Braque*, Penguin Books Ltd, 1959.

SCHWARTZ, P. W., *The Cubists*, Thames and Hudson, London, 1971.

WILSON, C., *The Outsider*, Victor Gollancz Ltd, 1956; Pan Books, 1963.

More About Penguins and Pelicans

Pelicans of related interest

The Psychology of Learning Mathematics
Richard Skemp

A book of interest both to those who are themselves studying mathematics and to those who are teaching it. Professor Skemp is convinced that to teach mathematics effectively, one has to understand how one learns.

Mathematics and Logic
Mark Kac and Stanislaw M. Ulam

Two eminent mathematicians explain to the layman how the mathematician thinks and what he is thinking about. 'Books which succeed in explaining the aims and scope of mathematics to the layman are as rare as hens' teeth . . . this is a very worthwhile book' – *New Scientist*

Mathematics and the Imagination
Edward Kasner and James Newman

A book packed with interesting puzzles and paradoxes, ranging from dice-throwing to rubber sheet geometry. The final chapters deal with the mathematics of change and growth and of the nature of mathematics itself.

Mathematics in Western Culture
Morris Kline

A masterly survey of the history of mathematical thought and the fascinating relationship between the nature of a society and its mathematics.

The Americans
Letters from America 1969–1979
Alistair Cooke

With his engaging blend of urbanity and charm, Alistair Cooke talks about Watergate and Christmas in Vermont, gives opinions on jogging and newspaper jargon, creates memorable cameos of Americans from Duke Ellington to Groucho Marx and discusses a host of other topics – all in that relaxed, anecdotal style which has placed him among our best-loved radio broadcasters.

'One of the most gifted and urbane essayists of the century, a supreme master' – Benny Green in the *Spectator*

The White Album
Joan Didion

In this scintillating epitaph to the sixties Joan Didion exposes the realities and mythologies of her native California – observing a panorama of subjects and ever ranging from Manson to bikers to Black Panthers to the Women's Movement to John Paul Getty's museum, the Hoover Dam and Hollywood.

'A richly worked tapestry of experiences' – Rachel Billington

The Seventies
Christopher Booker

From the rise of Mrs Thatcher to the murder of Lord Mountbatten, from the energy crisis to the trial of Jeremy Thorpe, from the Cult of Nostalgia to the Collapse of the Modern Movement in the Arts ... In this series of penetrating essays Christopher Booker explores the underlying themes which shaped our thoughts and our lives in the 'seventies.

'Booker is quite compulsive' – *Punch*

'Constantly stimulating ... savagely funny' – *Evening Standard*

Charmed Lives
Michael Korda

The story of Alexander Korda and the fabulous Korda film dynasty starring Garbo, Dietrich, Churchill and a cast of thousands.

'Charmed lives, doubly charmed book ... Comments, jokes, experiences; and at the heart of it all there is Alexander Korda, powerful, brilliant, extravagant, witty, charming. And fortunate: fortunate in his biographer. Few men have the luck to be written about with so personal an appreciation, so amused, yet so deep an affection' – Dilys Powell in *The Times*

Nancy Cunard
Anne Chisholm

The rebellious only child of an American society hostess and an English aristocrat, Nancy Cunard became the darling of high café society in the twenties and thirties. She knew Eliot, Joyce and Louis Aragon, she sat for Cecil Beaton and was an Aldous Huxley heroine in *Antic Hay*; a poet and a writer, she was the *avante-garde* publisher who 'discovered' Beckett, a passionate advocate of racial equality and a journalist in the Spanish Civil War. In her acclaimed biography, Anne Chisholm brings to life this extraordinary woman, Nancy Cunard, who, by the time of her tragic death in 1965, had become the dazzling symbol of her age.

An Actor and His Time
John Gielgud

The autobiography of one of the greatest actors of our time, told with wit wisdom and humour and – as the curtain rises on such famous names as Ellen Terry, Sarah Bernhardt and Richard Burton – vividly recapturing sixty golden years in the British theatre.

'A sparkling blend of reminiscence, anecdote, gossip and evocation' – Michael Billington in the *Guardian*

'The recollections hold the leaves of a benign autumn. Rustle them, and the century's theatre rises in unforgotten excitement' – J. C. Trewin in *The Times*

Recently published in Penguins

Carnival in Romans
A People's Uprising at Romans 1579–1580
Emmanuelle Le Roy Ladurie

'In February 1580, Carnival in Romans was a time of masks and massacres for the divided citizentry.' Concentrating on two colourful and bloody weeks, the author of *Montaillou* vividly resurrects the social and political events that led to the tragedy of 1580.

'Professor Le Roy Ladurie is one of the greatest historians of our time . . . this is a book not to be missed' – Christopher Hill

The View in Winter
Reflections on Old Age
Ronald Blythe

'Old age is not an emancipation from desire for most of us; that is a large part of its tragedy. The old want their professional status back or their looks . . . most of all they want to be wanted.' Ronald Blythe listened to all kinds of people who are in and around their eighties as they talked about their old age, to make this marvellous, haunting record of an experience that touches us all.

'Beautifully written . . . Moving but unsentimental and even oddly reassuring, it deserves, like *Akenfield*, to become a classic' – A. Alvarez in the *Observer*

The Old Patagonian Express
By Train through the Americas
Paul Theroux

From blizzard-stricken Boston to arid Patagonia; travelling by luxury express and squalid local trucks; sweating and shivering by turns as the temperature and altitude shot up and down; Paul Theroux's vivid pen clearly evokes the contrasts of a journey 'to the end of the line'.

'One of the most entrancing travel books written in our time' – C. P. Snow in the *Financial Times*